ONCE BEFORE TIME

ONCE
BEFORE
TIME

A WHOLE STORY
OF THE UNIVERSE

MARTIN BOJOWALD

ALFRED A. KNOPF NEW YORK 2010

Library of Congress Cataloging-in-Publication

Bojowald, Martin.

Once before time : a whole story of the universe / by Martin Bojowald.

p.   cm.

ISBN 978-0-307-27285-0

1. Cosmology.   2. Beginning.   3. Space and time.   I. Title.

QB981.B684 2010

523.1—dc22   2010015937

Manufactured in the United States of America

First American Edition

CONTENTS

PREFACE

... and if he does not do it solely for his own plea-
sure, he is not an artist at all.

—**OSCAR WILDE,** "The Soul of Man Under Socialism"

There are many reasons for a scientist to write a popular book, and many others not to. Research has primacy in science; this is where careers are forged and honors earned. Everything else wastes precious time—at least in the eyes of many a colleague who might one day be asked for an evaluation of one's work.

But what good does all scientific progress do if it cannot be communicated? Do we really understand the world if we cannot explain it without the requirement of long, demanding studies? Learning a complex matter too often means that we merely accept its crucial ingredients and principles, getting used to standard methods of calculation. A true test of our understanding comes only when knowledge is to be explained to an open-minded layperson free of preconceived notions. In this sense, quantum mechanics as one example is, despite all its many successes and technological applications, far from being understood (as indicated, perhaps, by the third chapter of this book). Writing a popular book is thus an exercise of utmost relevance for a scientist's own work.

A popular book is, moreover, the ideal place to allude to the unity of science, literature, and art. In all these areas one tries to picture the world and to communicate it. This unity, of course, does not exist in reality but only as an ideal. But a book that aims to be widely accessible has a right to call upon this ideal. For this reason, I am grateful to those who helped me tap this unity. In the realm of art, I thank Gianni Caravaggio, some of whose works are represented here and who contributed in several discussions to my understanding. Thanks also go to Rüdiger Vaas, who, over many years, has helped me to understand and to communicate my understanding. He was among the first to find my scientific results worthy of wide dissemination. Many others, who cannot all be named here, have continually forced me to leave the fortified ivory tower of science.

This book would not have come into existence without the original suggestion of Jörg Bong from S. Fischer Verlag, and the subsequent support of Alexander Roesler. For reading parts of my manuscript and for their many useful suggestions, I am grateful to Gisele Ben-Dor, Maryam Shaeri, and Hannah Williams. I thank the Physics Department of Pennsylvania State University, who know how to provide an exceptionally agreeable and stimulating atmosphere for its members. Early on, they offered me a free semester without even knowing about my writing plans! Penn State's Institute for Gravitation and the Cosmos has afforded me unique opportunities for multidisciplinary discussions and research related to topics in this book. The expertise of many of my colleagues has, at least subliminally, found its way into my writing.

I thank Elisabeth and Stefan Bojowald for their critical reading of an early version of the book, and for some hints such as those about cyclic images in Egyptology. In conceptualizing some passages, I was inspired by the tranquillity of their retreat at the edge of the Eifel Range.

STATE COLLEGE, PENNSYLVANIA

APRIL 2008 / SEPTEMBER 2009

ONCE BEFORE TIME

INTRODUCTION

The more abstract the truth you want to teach, the
more you must seduce the senses to it.

—FRIEDRICH NIETZSCHE, *Beyond Good and Evil*

The goal of science has always been nothing less than as complete an understanding as possible of the laws of the world—nothing less than as unequivocal a description as possible of what we see and probe. Nothing less than coming as close as possible to what can be considered truth—in a nonsubjective way, the only way that counts.

Over the course of the last century, physical research in particular has progressed far to build a dominant theoretical edifice: quantum mechanics and general relativity. Understanding nature on the large and the small scale has become possible, from the whole universe in cosmology all the way down to single molecules, atoms, and even elementary particles by means of quantum theory. Precise descriptions and a deep understanding of a wide variety of phenomena have resulted, and they have been spectacularly confirmed by observations. Especially during the past decade, this hallmark of scientific success has also been achieved in the cosmology of the early universe.

Aside from its technological relevance in almost all areas of everyday life, an unmistakable sign of the quality of scientific progress is that for quite some time, some fields of scientific inquiry have touched upon questions traditionally held to be in the realm of philosophy (giving rise to the term "experimental metaphysics," coined by the physicist and philosopher Abner Shimony). Since Aristotle, the aim of all theory has been to shed light on general phenomena and to understand their causes, in contrast to collecting disconnected bits of knowledge. Philosophy, by contrast, asks for the deepest origin or principles of all that exists. In this sense, the merging of some physical and philosophical issues can in fact be considered a distinguishing feature of scientific progress. When physics pushes ahead to such questions, it gains a position that allows it to contribute to discussions of far more general—and more far-reaching—interest. In the context of a combination of cosmology and quantum theory, the most important question is that of the emergence and the earliest phases of the universe, a question that has preoccupied humankind ever since the beginning of philosophy—and even before.

Other questions that have engaged thinkers over the centuries, and that remain of great significance, in quantum theory as well as general relativity, are the role of observers in the world and the question of what can be observed at all and what perhaps cannot. In cosmology, the entrance of physical research methods means the emergence of empirically testable scenarios for the whole world. The big bang model is founded on general relativity—as a description of space, time, and the driving gravitational force—as well as on quantum theory, which is indispensable to understanding properties of matter in the early universe. From all this, a breathtaking explanation results for the successive emergence of all matter—nuclei, atoms, and compound material objects all the way up to galaxies— out of an extremely hot initial phase.

At this rarefied place, the limits of the established worldview become visible. In spite of all their successes, general relativity together with quantum theory, as they are being used today, do not provide a complete description of the universe. When one solves the mathematical equations of general relativity in hopes of finding a model for the temporal evolution of the universe and its long-term history, one always reaches a point—the so-called *big bang singularity*—where the temperature of the universe was infinite. It is no surprise that the universe was very hot in the big bang phase; the expanding universe was, after all, much smaller and more densely compressed at those times than it is now, implying an enormous increase in temperature. But infinity as the result of a physical theory simply means that the theory has been stretched beyond its limits; its equations lose all meaning at such a place. In the case of the big bang model, one should not misunderstand the breakdown of equations as the prediction of a beginning of the world, even though it is often presented in this way. A point in time at which a mathematical equation results in an infinite value is not the beginning (nor the end) of time; it is rather a place where the theory shows its limitations. In spite of all its successes in other areas, the theory given by general relativity in combination with the quantum theory of matter remains to be extended.

The problem lies in the incompleteness of the revolution brought about by physical research during the last century. Quantum theory is used to describe matter in the universe, but not gravity or even space and time themselves. The latter are firmly in the domain of general relativity, largely independent of quantum physics. A successful combination of quantum theory and general relativity, even in the realms of space and time, would significantly extend known theories. Such a combination—*quantum gravity*—is particularly important for a description of the hot big bang phase of the universe, and it can explain, or so we hope, what happened at the infinity of the big bang singularity. Was this really the origin of the world and of time, or was there something before? And if there was something before the big bang, what was it?

Unfortunately, quantum gravity is extremely complicated. Even separately, general relativity and quantum theory are distinguished by a mathematical machinery unknown to preceding branches of physics. To make matters worse, these two areas require markedly different mathematical methods. A combination of the physical theories also requires a unification of the underlying mathematical principles, further magnifying their degree of difficulty. For this reason, no completely formulated quantum theory of gravity is available yet, in spite of many decades of research and vigorous efforts by a large number of scientists. What we have seen nonetheless, in particular in the last few years, are numerous promising indications for some properties of quantum gravity that can already be analyzed. As so often in research, the situation resembles the first stages of assembling a jigsaw puzzle, when one may already have an inkling of the emerging picture but could well be on the wrong track. Our current view indicates what a completion of physical theory can bring about: It allows us to see what possibly happened at and even before the big bang. We are granted a glimpse of the earliest times of our universe and can, for the first time, analyze how it may have arisen.

In this book, recent results of the theory as well as plans for satellite observations in the near future are explained, and it is shown how radically they could alter our worldview. *Loop quantum gravity* in particular, one of the variants currently put forward as a possible combination of general relativity and quantum theory, has provided first results concerning a nonsingular description of the big bang. In this framework, the universe existed before the big bang, and one can roughly estimate how it could have differed from what we see now. By its influence on later stages of the cosmic expansion, detectable by sensitive observations, this ancient prehistory of the universe can be explored. This book provides a firsthand report of this line of research, followed by a discussion of black holes, which also show fascinating effects. The final chapters touch upon further issues regarding a general understanding of the world, among them cosmogony, the riddle of time and its direction, and the grail of a "theory of everything." In parallel with progress in our scientific understanding, the human trail of knowledge will itself, from a personal perspective, be illuminated by examples of modern research.

Although the required theory is highly mathematical, many calculations are by now intuitively understood. Intuition is not only useful for explorations in an unknown territory; it also allows broad explanations. As suggested by the Nietzsche passage at the beginning of this Introduction, the aims of this book may be realized without the use of mathematical formalism, except for a single crucial equation in chapter 4. While mathematics cannot be dispensed with if one wishes to discover and fully understand the issues underlying this book, an intuitive grasp is possible without too much effort. One may not always discern why things are supposed to be one way rather than another, but with some trust in the travel guide, connections can be grasped.

A warning is required nonetheless: Many areas of research in quantum gravity are still to be considered speculative. In contrast to the first half of the past century, the period when general relativity and quantum theory were developed, there are as yet no observations to corroborate the theoretical formulations of quantum gravity. What currently propels research is a motley collection of conceptual considerations regarding the incompleteness of general relativity, as well as mathematical consistency conditions in the formulation of equations. For instance, there is no guarantee whatso-

1. The philosopher's stone melts away. What counts as certain knowledge may, upon further examination, turn out to be in need of correction. Evaluating the results or promises of science must always take into account its limits. Often, such limits are even more important than established results, for they show the way to new insights. (Sculpture by Gianni Caravaggio: *Spreco di energia assoluta* [*Absolute waste of energy*], 2006. Photograph: Robert Marossi.)

ever that a combination of certain mathematical methods, as they are used in general relativity and quantum physics, will allow solutions for a reliable description of the universe. The required mathematical tools are indeed so restrictive that formulating a theory with reasonable solutions would in itself represent immense success. Whether there may be other reasonable theories is a different, so far incompletely understood question. This shows the fragility of the pillars on which quantum gravity presently rests. But optimism does prevail, for many independent indications such as those in this book point in the same direction. What's more, it is expected that cosmological observations made in the not too distant future could reveal phenomena predicted by quantum gravity. Such potential observations, which are also described in this book, would finally render quantum gravity an empirically tested theory.

As of now, the status of quantum gravity resembles the earliest stages of claiming a new territory. Mathematics is the pioneer who opens up new areas beyond established frontiers. In our case, these frontiers are literally those of the universe and of time. Mathematics also serves to explore newly won lands; but in an empirical science such as physics, these lands can be claimed and secured only by observations. This remains to be accomplished for quantum gravity, which thus resembles territory still full of danger, where one can all too easily get lost or be swallowed up in swamps of speculation.

Such a land demands a humility toward nature that is not always shown. The language of physicists often sounds very determined (and sometimes even shows signs of hubris), but concerning a law of nature one must bear in mind, "It may be accurate, it may be faulty. If it is not accurate, the scientist, not nature, is to blame."[1] A physicist constructs laws of nature, but is the one responsible if they are broken. Nobody is a physicist's subject, least of all Nature herself. This is true in particular for theoretical sketches such as quantum gravity. In the meantime, before observations can show that Nature pays at least some respect to the laws imposed on her, Intuition will be the guide in this unknown land—on an adventurous trip back before the big bang.

1. GRAVITATION
MASS ATTRACTION

Should something from the window fall
(and if it just the smallest be)
how jumps the law of gravity
as mighty as wind from the sea
at every ball or blueberry
and takes them to the core of all.

—RAINER MARIA RILKE, *The Book of Hours*

Over large distances, the universe is governed by the gravitational force. In physics, the action of a force is the cause of motion or of any form of change. Complete rest is possible only if no net forces are acting. One scenario in which this can happen is the absence of any matter whatsoever—a state called vacuum. But matter quite obviously does exist, and just by its mass it causes gravitational forces on other masses. To realize motionless states of rest, at least approximately, all acting forces must compensate each other. In addition to gravity, there are the electric and magnetic forces to be considered, as well as two kinds of forces called weak and strong interactions, reigning in the realm of elementary particles.

While the electric force is easily compensated over large distances by the existence of positive and negative charges, mutually neutralizing each other, the forces that come into play in the interior of nuclei act only at extremely short range. What remains over long distances is gravity alone. It rules the general attraction of masses and energy distributions in space, and thus dictates the behavior of the universe itself. In contrast to electricity, there are no negative masses: Gravitational attraction cannot be fully compensated. Once massive objects such as stars or entire galaxies form, the resulting gravitational interaction dominates all that happens. The facets of this commonplace force, often ignored in recent research and yet—in cosmology and black holes—giving rise to a rich variety of exotic phenomena, are the topic of this book.

NEWTON'S LAW OF GRAVITY:
DISTANT ACTION AND A FATAL FLAW

The first general law of gravity was formulated by Isaac Newton. As is typical for many important steps in gravitational research, this theoretical development required a unified view on well-known phenomena on Earth with a long list of intricate observations of objects in space: the moon and some planets. The latter was accomplished thanks to technologies that, for those times, were highly sophisticated; conversely, such research has spawned the development of new instruments. Combining fundamental questions and technological applications, in many areas of science and in gravitational research in particular, is a success story that continues into the present day.

Even before Newton, the initial untidy flood of data, as it was accumulated by astronomers such as Tycho Brahe, Johannes Kepler, and many others, was ordered into a model of the solar system. Since Nicolaus Copernicus and Kepler, this model has largely held the form we know today: Planets orbit around the sun along trajectories that, by a good approximation, can be considered as ellipses, or slightly oblong circles. But what is propelling the planets along their curved tracks? From common observations we know that a force is necessary to keep a body from moving stubbornly along a straight line. How can one describe or even explain the required force in the case of the planets?

Newton's groundbreaking insight—the existence of a universal force of gravity causing not only the motion of all planets around the sun, and of the moon around the earth, but also the everyday phenomena of falling objects—is impressive. It is an excellent example of the origin of scientific explanation: not an answer to a "why" question in the sense of an anthropomorphic motivation, but a plethora of complicated phenomena, unrelated at first sight, reduced to a single mechanism: a law of nature. Newton's mathematical description of the situation is very compact and highly efficient for predictions of new phenomena described by the same law. In the case of Newton's law of gravity, the unfathomable power of theoretical prediction has repeatedly been employed—for instance, to find new planets via small deviations imposed by their gravitational pull on the trajectories of other planets, or in planning modern satellite missions.

Such success stories, in which an elegant mathematical description explains and predicts a multitude of phenomena, can be found throughout physics; they are indeed the landmarks of its progress. Reliving such insights is often so gratifying that scientists employ the term "beauty"—a pragmatic kind of beauty whose core, the mathematical formulation, can be seen only by the initiated, but which in its concrete successes can also be appreciated by outsiders.[1]

Concretely, Newton's law of gravity describes the attractive force between two bodies caused by their masses. The force increases proportionally with the amounts of the masses: The attraction between two heavy bodies is larger than that between two light ones. It is also inversely proportional to the squared distance between the bodies; it weakens considerably when the bodies are farther apart. In addition to these proportionalities, the exact quantitative strength of the force is determined by a mathematical parameter, now called *Newton's gravitational constant.* In this value one can see the unification of earthly and heavenly phenomena. The gravitational constant can be derived from the tiny attraction of two masses on Earth, as was first accomplished in Henry Cavendish's laboratory in 1797 and '98; using the same value to calculate the force exerted by the sun on the planets shows exactly the right nudge required to hold the planets on their observed orbits.

In contrast to its clear dependence on distance, Newton's gravitational force is completely independent of time. Time independence sounds plausible, for a fundamental law of nature should, after all, be valid at all times in the same way. It is also consistent with the dominant understanding of space and time in Newton's age and long thereafter, not to mention our everyday conceptions of them. Although one can easily change the positions and distances of objects in space, space itself appears unchangeable. Also, time seems to pass simply and uniformly, without being influenced by physical processes or technical instruments. Since gravity, according to Newton, acts instantaneously—independently of how far apart the masses are—the force need be formulated only for the case of two masses not at the same place, but at the same time.

Despite its plausible form and celebrated successes, Newton's theory did have a flaw in its beauty. Like the beauty of the theory itself, this flaw, too, can be understood completely only with a sufficient amount of background knowledge. But even on the surface, it is a good example of the progress of theoretical physics. Newton himself had reportedly been uneasy about the "animalistic" tendencies of his law of gravitation: As an animal is attracted from far away by the expectation of food or companionship, a massive body appeared to move toward another one from a distance. This action at a distance, apparently without the more intuitive type of local interactions as realized for bodies pushing each other at close contact, was considered a serious conceptual weakness in spite of all concrete successes.

It is extremely difficult to correct this weak spot by constructing a theory only of local interactions that, of course, should otherwise remain compatible with the astronomical successes of Newton's theory. To start with, one will have to consider the time dimension, too, for such a local interaction must take some time to propagate from one body to the other. As it turned out, a consistent reformulation is possible only by radically changing Newton's—and our—intuitive conceptions of space and time. It requires much more highly sophisticated mathematical machineries and substantial efforts, but these efforts are rewarded by a theory of unprecedented beauty in the sense described above. All this required dedicated physical research and, not least, a strong mathematical grounding. The flaw in Newton's theory was to be corrected only long after Newton—by Albert Einstein.

RELATIVITY OF SPACE AND TIME:
SPACE-TIME TRANSFORMERS

All this took a long time, or a short time: for, strictly speaking, *no* time on earth exists for such things.

—FRIEDRICH NIETZSCHE, *Thus Spoke Zarathustra*

In physics, as in all of science, it is important to distinguish between properties that depend on the person making an observation and properties independent of an observer. The mass of a particle refers only to the particle itself and will, if the particle remains unchanged, always be measured as the same value. Except for unavoidable experimental inaccuracies, it does not matter who is doing the measurement. A particle's velocity, on the other hand, appears different, and sometimes drastically so, depending on whether an observer is moving with respect to the particle. An observer moving along with the particle at exactly the same speed would perceive the particle as being at rest, well known from two cars cruising side by side along a straight stretch of highway. To the driver of one car, the other one seems not to be moving. Any other observer would see the car (or the particle) move and attribute to it a nonzero velocity. Relativity in general terms is the mathematical analysis of such relationships; it ultimately tells us what we can learn about nature in a fully objective, observer-independent way.

For many centuries, space and time were thought of as observer-independent. Distances between points and durations of temporal periods appeared absolute, no matter how an observer would be positioned or move. But the first fault lines in this worldview opened up toward the end of the nineteenth century, eventually leading to special relativity. In this new view, space and time cannot be seen in separation but are intertwined, interchangeable, and observer-dependent. Like the velocity of a particle, the values measured for them depend on the motion of an observer. In abstract terms, they describe different dimensions of a single physical object: space-time; and only space-time concepts, but not space or time themselves, are independent of the person making a measurement.

How can this be demonstrated by physical means? To answer this question and to explain the role of dimensions, we first consider space alone. Space has different dimensions, namely three: we can move sideways, back and forth, and up or down. Here, one might ask why these should be considered as three dimensions of a single space, rather than three completely independent directions: width, depth, height. The answer is simple. Width, depth, and height are not absolute and independent properties; they can be commuted into one another. We have only to turn around in space to make the height of a cube appear as its width, and in this sense height and width can be interchanged. This is not a transformation by a physical process, like a chemical reaction, but a much simpler one by means of changing our viewpoint. What we see as height, width, and depth depend on the place of an observer (or on conventions such as the use of Earth's surface along which to measure width and depth); they cannot be considered properties of space itself as a physical object. For this reason, one speaks of three-dimensional space, not of the existence of three independent one-dimensional directions.

Time is similar, although its transformation is harder. By simply turning around one can influence only one's view on space; the change of the angle of view (or, more precisely, the tangent of the angle as a mathematical function, which does not differ much from the angle when it is small) is expressed by the ratio of spatial extensions, such as the height before and after changing the viewpoint. By changing the angle, one can only transform spatial extensions into one another. If we want to transform space into time, we must vary a quantity given by a ratio of spatial and time extensions: a velocity. Traversing a certain distance in some period of time means that one moves at a velocity obtained as the ratio of that distance to the required time.

This consideration does, in fact, lead to the basic phenomenon of special relativity. If we are moving faster than a second observer while viewing a certain scene, spatial and time distances appear different to each of us. As changing the angle of view transforms spatial extensions, changing the velocity of an observer commutes spatial distances to timelike ones and vice versa. Distinguishing space and time extensions is thus dependent on the viewpoint (or the "viewtrack," if we are indeed moving); it cannot have a physical basis independent of observers' properties. Instead of separate space and time, there is only one joint object: space-time. Special relativity is the theory of these changing viewtracks (also called *inertial observers*) in the absence of the gravitational force.

As an illustration, these considerations are certainly no proof; not every ratio implies a transformation when it is changed. For instance, the birthrate of a country is the ratio of newborns to the total population, but a change in the birthrate does not mean that inhabitants are transformed into newborns. An important difference from the previous example is the role of observers: Changes are caused by observers taking different positions and states of motion; and since physical laws must be independent of the special private and personal properties of those making the observations, concepts distinguished only by viewpoints must be discarded. In special relativity, this "transformability" of space and time, forcing us to deny them separate meaning, has not only been substantiated mathematically; it has also been verified experimentally myriad times, especially in reactions of elementary particles. While the Newtonian concepts of a rigid space and an independent time would not agree with many measurements made in the last century, in a special relativistic view no inconsistencies arise.

Newton's view was able to enjoy great success for such a long time because noticeably transforming space and time requires very large observer velocities. Unless measurements are extremely refined and precise, in order to see an effect, speeds must be close to the immense velocity of light: roughly 300,000 kilometers per second. In everyday life, this makes the transformability of space and time imperceptible.[2] For an observational verification, one needs either very high velocities or very precise time measurements in order to notice the tiny time changes at low velocities. Both methods have been developed in the past century.

Very precise time measurements are achieved by atomic clocks, making space-time transformations detectable even at the typical speeds of airplanes. (Since planes have to move at a certain height, additional effects arise due to a reduction of gravity acting on the clock farther away from the center of the earth. This general relativistic effect, depending on the gravitational force, is introduced below.)

At velocities close to that of light, space-time changes drastically: As an observer at rest would describe it, time is transformed almost completely into space, and thus passes ever more slowly. Once the speed of light is reached, which is possible only for massless objects such as light itself, all timelike distances vanish. Going beyond that speed limit is impossible, for all time has already been used up when we reach the speed of light. No signal can move faster than light. Delays in any transmission of information must always occur; they may be small, but they do become noticeable at large distances. (This maximum speed is that of light in a vacuum. In transparent media such as water, light usually moves more slowly than in a vacuum. There are thus signals in such media that propagate faster than light in the same medium, but not faster than light in a vacuum.)

High velocities can be probed, too, although not by violently accelerating a clock, but by employing fast clocks that nature provides. Earth is bombarded from outer space by highly energetic particles moving at nearly the speed of light.[3] Most of them do not reach the ground, because they react with oxygen and nitrogen nuclei in the upper atmosphere and produce new particles, among them the *muons.* Muons are a heavy kind of electron and quite similar to them, except that they are unstable: A muon at rest decays after just a millionth of a second, producing an electron and two other stable particles called *neutrinos.* The decay time may be used as a unit of time telling us when a millionth of a second has passed at the speed of a muon. By the standards of modern technology, this "muon clock" is not very precise. But muons can be brought to high velocities much more easily than atomic clocks, and a free supply of them is showered down on Earth in a quickly moving state every day in the form of natural cosmic radiation.

Cosmic muons lead to an impressive confirmation of special relativity and its transformability of space and time. Even racing at such high velocities as they acquire in the upper atmosphere, the muons' lifetime of a millionth of a second would not be nearly enough for them to reach the ground. And yet detectors do register many of these particles, even though they are supposed to have decayed along the way. The resolution of this problem is that a millionth of a second, within which a muon at rest would decay, seems much longer for a muon traveling at high speed and watched from the ground. Thanks to the high muon velocity, from the particles' viewpoint enough space is transformed into time for them to reach the ground before their decay.

Measurements with atomic clocks or of muons were not available when Einstein developed his theory of special relativity. Rather, he derived his equations for the transformation of space and time from a deep consideration of the theory of light, introduced in 1861 by James Clerk Maxwell. (As a mathematical curiosity, the same transformation law, but without a physical interpretation, had already been found by Hendrik Antoon Lorentz.) The process of applying such principles independently of observations can be compared with Newton's realization of his own theory's incompleteness. Newton's law of gravity was, at its inception and long thereafter, highly successful in the description of astronomical observations; it was centuries before the first minute and unexplained deviations from the law were observed. Even so, as mentioned above, Newton was not completely happy, for his law looked too animalistic. What makes two masses attract each other even when they may be arbitrarily far apart? This flaw, already looked upon with suspicion by Newton, becomes acute in special relativity.

In Newton's understanding of separate space and time, there is no problem in principle with his gravitational law; there may at best be an aesthetic one. In special relativity, however, the law flatly becomes inconsistent. Newton's gravitational force depends on the distance between two bodies in space, but there is no quantity related to time. If one tries to combine this with a transformability of space and time, a strict application of the law would mean that the gravitational force had to depend on the state of motion, e.g., the velocity of a measurement apparatus: A change in the velocity must transform space in time, causing Newton's law to be time-dependent. The smaller spatial distance would then be compensated by the larger time lapse in such a way that all observers, moving at different speeds, would calculate the correct force. But this requirement was not taken into account when Newton formulated his law; thus the need to extend Newton's theory.

A similar situation is realized in the theory of electromagnetism.

Coulomb's law for the electrostatic attraction (or repulsion) of two charged bodies, so called after its formulator, Charles Augustin de Coulomb, is very similar to Newton's for the gravitational attraction of two masses. One simply has to replace masses with charges and Newton's gravitational constant with an analogous one to quantify the electric force. (Moreover, one has to change the sign of the force, indicating its direction: Two charges of equal sign repel each other, while two—always positive—masses attract each other.) In particular, the dependence on the distance is the same in both cases, and in none of these formulas does a distance in time appear. For Coulomb's law, an extension had already been found by Maxwell for different reasons, based on the relationship between electric and magnetic phenomena. This extension turned out to be consistent with the transformability of space and time; and it had been found well before Einstein, for whom it played a major role in his considerations leading to special relativity. Maxwell, however, did not recognize the connection between his extension of Coulomb's law and the changes of space and time.

In 1905, when Einstein developed special relativity, uncovering the observer-dependent form of space and time, no reformulation of Newton's law existed. The necessary extension of the theory, addressed afterward by Einstein himself, proved to be much more complicated than Maxwell's generalization of Coulomb's law. Another decade had to pass before Einstein came up with its final form—general relativity—in 1915. The payoff was not only a gravitational law reconciled with the principles of special relativity, but also a further radical change in our understanding of space and time as well as a mathematical foundation for cosmology. In the remainder of this chapter we will primarily be occupied with the structure of space and time, and much later come back to its role in the behavior of the whole universe, in the chapter on cosmogony (chapter 8).

GENERAL RELATIVITY:
SPACE AND TIME UNBOUND

Effortlessly swings he the world, by his knowing and
willing alone.

—XENOPHANES OF KOLOPHON, Fragment

Special relativity does not include the gravitational force; in this
sense it is not general enough. With general relativity a gravitational law is available that is consistent with special relativity. But it
is not only an extended, more contrived form of Newton's law; the
theory of general relativity constitutes nothing less than the final
promotion of space-time to an object of physical meaning. What is
considered space and what is considered time is not only changeable,
as in special relativity, depending on the viewpoint of an observer; it
is itself dynamic and subject to physical processes: The form of
space-time is determined by the matter it contains. Space-time is not
a straight, flat, four-dimensional hypercube extending unchanged all
the way to infinity. Like a piece of old rubber, it writhes under its
own inner tensions into a curved structure. The inner tension of
space-time is the gravitational force.

As velocities have to be very large to clearly bring out the effects
of special relativity, changing space and time just by having a different viewpoint, the influence of gravity on space-time is usually
weak. By currently available means it is impossible to exploit this
technologically (even though we sometimes speculate about the construction of wormholes, warp drives, or mini–black holes). In astrophysics or cosmology, however, there are often objects so heavy that
a precise description must take into account not only their matter
content but also properties of space and time themselves. This has

led to many tests of general relativity along with, as will be detailed later, new worldviews in cosmology.

In 1915, Einstein did not have such observations at his disposal, just as his earlier formulation of special relativity was driven by thought rather than experiments; his constructions were based solely, but solidly, on the possibility of a mathematically consistent realization of his overarching principles. The result is a theory whose elegance remains unsurpassed in physics. Drawing on general principles and a geometrical form of mathematics, which through a long and noble pedigree can be traced back to its eminent beginnings in ancient Greece—in the case of geometry, to Plato and Euclid— one must almost inevitably arrive at a unique form of equations describing the entire cosmos.

Einstein had to struggle for a long time before he understood the right principles and the required mathematics, but his work was eventually crowned with immense success. Not only did the theory satisfy the highest demands of mathematics, a field in which it continues to provide important stimulus for further research, but later it was also able to explain many observations that Newton's theory could not.

Such long-lasting implications surely justify the great interest in Einstein's work; but in the last decades, alas, this success has often turned into a curse. In wide circles of physicists, the prevailing opinion often seems to be that general relativity is already completely understood and experimentally fully verified. Sometimes such a view is even used to justify cutting back research, and with it jobs, in this area. A complete verification of a theory is in any case never possible, and for this reason alone we should never forgo new experiments that might provide independent comparisons of theory and observations—especially with such an important and fundamental idea as general relativity. The set of experiments testing a theory can, at any given time, cover only a limited range of phenomena. An experimentally tested theory may be successful to a certain degree, but we can never be certain that it correctly describes all processes to which it can in principle be applied. Just as Newton's theory was consistent with observations for a long time, until it was recognized as a special case of general relativity with a limited range of validity, general relativity, too, could turn out to be a limiting case of an unknown theory yet to be found. Even on a purely theoretical basis, relativity remains incompletely understood; there are many unanswered questions, of direct importance in particular for cosmology, and thus an acute need for research remains. There is indeed mounting evidence that general relativity itself needs to be extended.

Most physical theories are established through a long and tedious process starting from a creative idea, or in other cases from an observation not explicable by available knowledge. An idea may be followed up on because it might appear attractive from an aesthetic or mathematical perspective; a new, unexplained experimental result might force us to change current theories so that they agree with the new observation. Such a process can go on for decades, and it keeps scores of physicists busy—theoretical as well as experimental ones. Many currently hot theories, such as those of particle physics or quantum gravity, are still subjected to this process. The development of quantum mechanics also followed this course for a long time, until it was cast in its currently accepted form. (Even here, many foundational questions remain open; from the perspective of physical applications, however, quantum mechanics can be said to be understood.) The end result, as it enters textbooks later on, is often hardly recognizable compared with the early formulations; many a historical contribution has turned out to be unimportant, too complicated, or just plain wrong. For theories still in development, it is not even clear whether they will ever become a solid part of our worldview; entire branches of physics can come to a dead end, even though research always provides lessons that then become important elsewhere.

The situation was completely different when Einstein devised general relativity. Einstein alone, supported only by a few friends such as Marcel Grossmann and in a certain competition with the mathematician David Hilbert, provided the decisive work. Not all contributions followed a direct step-by-step route, and some of the published articles did turn out to be fruitless. But in a relatively short amount of time he completed his work, which soon proved successful in confrontations with observations. One may easily get the impression that Einstein had immediately created his theory in perfect form, without the need for a lengthy series of studies and improvements. Certainly, this impression can explain why even some physicists no longer consider general relativity worthy of new research.[4]

In reality, however, the picture is different. Only the simplest solutions of general relativity are understood, which, fortunately, suffice for many questions in physics; even the simplest and most highly symmetric solutions afford impressive insights into cosmology and astrophysical objects such as black holes. But if one tries to go only one step in complexity beyond such solutions, one encounters immense difficulties owing to the complicated form of the theory. Its equations are of a type allowing hardly any standard solution procedures to be applied. Every situation has to be analyzed anew, and only in a few lucky cases can exact solutions be found. Using computers sometimes helps, but even then the equations resist easy analysis. For these reasons, many mathematicians are interested in several issues of general relativity, and time and again they have contributed to our understanding of it. Open questions also exist, such as that of predictability (see chapter 6), that have a bearing on the foundation of physics as a whole.

A numerical analysis of Einstein's equations—often the last hope when direct mathematical solutions turn out to be too complicated—is extremely difficult. Computational research was begun in the 1970s and received considerable support in the 1990s. Collisions of heavy stars or black holes were of particular interest because they were expected to be strong sources of an entirely new kind of signal: gravitational waves. General relativity predicts that space-time itself can be excited to vibrations, periodic ripples that then propagate in the form of waves just like those on the sea. We have strong hopes of seeing these in coming years with sensitive detectors; such developments would not only further test general relativity but also open up a new branch of astronomy. We could start to explore the cosmos not only by light or other forms of electromagnetic radiation, but also with the help of gravitational waves. It is as if we would be able not only to glance into the sky, but also to listen to it. A new sense would be opened, enabling and ennobling unprecedented experiences and insights.

To detect gravitational waves, like any new signal, one has to know what to look for: One must know the intensity of a gravitational wave as it changes in time while traveling to us, much like a wave through water, starting from its creation in a collision. Unfortunately, the mathematical equations are too complicated for a direct solution, and even computers have been of little use; frustratingly, the programs available crashed before they could show interesting results—it was like having to type a long text in a program that would quit after entering every single word. Only after intensive activity over many years, performed in several groups (whose total number is still small compared to collaborations in particle or condensed matter physics), did a breakthrough recently become possible. As first shown in the work of Frans Pretorius in 2005, stable computer programs can now be developed that are able to provide valuable insights into heavy object collision results. Just in time, the construction of gravitational wave detectors such as LIGO, a set of detectors in the states of Louisiana and Washington, GEO600 in Germany, Virgo in Italy, and TAMA in Japan is rapidly under way; the dream of gravitational wave astronomy may soon become a reality. None of this would be possible without the foundation of general relativity, as buttressed ever more firmly by continuing research.

But back to the historical developments. Einstein was certainly not working completely independently of observations, for he sought to extend Newton's astronomically tested gravitational law. Grounding in established laws is important for any kind of progress in physics. But Einstein had little further experimental guidance as to how Newton's theory was to be extended. There were merely tiny observed deviations in some planetary trajectories, in particular in Mercury's orbit, which, as noticed first by Urbain Le Verrier in 1855, seemed to shift compared to Newtonian calculations by the small amount of 43 arc seconds (about one one-hundredth of a degree) per century. The influence of Venus, the planet closest to Mercury, was already factored in, as were small perturbations such as possible irregularities in the shape of the sun, without managing to reconcile the observations with theoretical considerations. Only Einstein was able to explain the shifted trajectory in a natural way by using his new equations of motion in general relativity.

Fortunately, new data soon arrived that were also incompatible with Newton's law but that had already been predicted correctly by Einstein. These were tiny shifts, or bending, experienced by starlight in close passage around the sun. Measured by Arthur Eddington during a total solar eclipse in 1919, these shifts led to the first triumphant verification of general relativity. (By now, more precise measurements of this type have been performed using radio waves emitted by quasars, as done for the first time by Edward Fomalont and Richard Sramek in 1976.) If deviations between Einstein's predictions and those observations had occurred, Einstein's theory would have been long forgotten, despite his quip "If nature does not coincide with theory, it is all the worse for nature" ("Wenn die Natur nicht mit der Theorie übereinstimmt, so ist dies um so schlimmer für die Natur").

In 1960, Robert Pound and Glen Rebka performed the first test of general relativity in an experiment on earth, and this test, too, was passed with highest marks. Here, the transformation of time at different altitudes, implying different positions in space-time, was measured. Farther away from the center of the earth, the gravitational force is weaker, which geometrically implies, as we will soon see, a changed form of space-time. Time progresses differently at higher altitudes, becoming faster than at lower ones. This speed-up is normally not noticeable, but it can be detected by sensitive measurements. To that end, Pound and Rebka exploited the Mössbauer effect, endowing some crystals with very finely tuned frequencies for the emission and absorption of light. Matter such as an atom can usually emit and absorb light near certain frequencies in the so-called spectrum, as it is used in fluorescent lights or lasers. The reason for the select set of frequencies is the quantum nature of matter (the topic of the next chapter). Since single atoms or molecules, on which such measurements would be performed, move in a gas, emission and absorption processes occur in different states of motion; the atoms, after all, move due to heat. Emission and absorption take place at different velocities; and according to special relativity, the progress of time, thus the frequency as the number of oscillations per unit of time, depends on the state of motion. Therefore, light is emitted and absorbed not at fixed and precise frequencies, but in frequency intervals of finite width.

In bodies subject to the Mössbauer effect, emission and absorption happen not at single atoms but at the entire crystal. As a whole, the crystal moves less than atoms in a gas. Accordingly, emission or absorption frequencies are much more precise. Special relativity no longer implies deviations of frequencies; but when a light-emitting crystal and an absorbing one of the same type are positioned at different altitudes, general relativity comes into play. Time progresses differently for the emitting crystal than it does for the absorbing one, causing a frequency mismatch in the light that reaches the absorbing crystal. A great enough mismatch prevents the light from being absorbed, and this is exactly what one can detect without even using large differences in altitude: The height of a building of several stories is sufficient.

The same relativistic phenomenon, using not the Mössbauer effect but the precision of atomic clocks, was observed in 1971 by J. C. Hafele and Richard Keating through sensitive comparisons of the progress of time aboard airplanes. In this case, it is special relativity that is important owing to flight velocity, as well as general relativity owing to altitude, as the airplanes changed positions along the gravitational force. Even so, the importance of general relativity was not always recognized even long after this successful experiment. On June 23, 1977, the satellite NTS-2 was launched, the first to carry an atomic clock for experimental purposes. The atomic clock was constructed so as to correct for the relativistic changes in the flow of time. But the developers of the satellite were not fully convinced of the need for corrections for general relativity; instead, the clock was equipped with a gadget to shift the clock rate to the correct value if necessary. After about twenty days in space, the signals indeed showed a deviation in the clock rate compared to clocks on Earth, exactly as predicted by general relativity. In this case, fortunately, the mistake could be corrected by switching on the frequency shift.

Perhaps the most impressive confirmation of general relativity was achieved through observations of double pulsars. These are systems of two stars closely orbiting around each other, one of which (the pulsar) emits radiation at regular intervals. The reason for the emission is usually a rapidly rotating neutron star that, like a lighthouse, emits signals into space and to us. Depending on the position of the pulsar in the double system, signals are delayed at different rates, for they must traverse different routes to reach us. The orbit and possible changes of the system's radius can thus be determined very precisely. In particular, general relativity predicts that gravitational waves are emitted during the orbiting process, causing the system to lose energy. The loss of energy implies that the two stars approach each other, which should be detectable by precisely measuring the orbit. The loss would be largest when the stars are closest. Each of them then sits more deeply in the gravitational field of its partner and general relativistic effects are stronger.

In 1974, Joseph Taylor and Russell Hulse identified a very close double pulsar consisting of two 20-kilometer-radius neutron stars orbiting around each other in just one-third of a day.[5] Their closest distance from each other is a mere 700,000 kilometers, roughly the radius of the sun! This is an ideal test system for slight deviations in the orbit as they are predicted by general relativity, and indeed, observations still being made up to the present day agree exactly with the predictions. (This system is, by the way, also subject to an additional shift in the orbit just like Mercury's, which does not involve a change of the radius as the energy loss due to gravitational waves does. Rather, one can visualize it as a slowly rotating oval whose shape shows the planet's orbit and its shift. In the double pulsar, the angular shift is four degrees per year—far larger than Mercury's—and can be used to estimate the neutron stars' masses.) Since then, many more close double pulsars with a wide range of orbital properties have been discovered, allowing a large variety of observational tests.

One of the most recent experiments is Gravity Probe B—a satellite launched on April 20, 2004, to collect data for sixteen months in orbit. The idea for this mission was first conceived in 1959, but developing the required technology took a long and tedious road even under the skillful direction of Francis Everitt. The effects aimed at—namely, a "yanking" exerted on space-time in the vicinity of the rotating Earth—demand extremely precise gyroscopes. To avoid perturbations by too uneven a shape, which would prevent any measurement of such sensitive effects, the gyros had to contain the most perfect spheres ever constructed. Even the whole universe can offer few rivals; only some very dense neutron stars are smoother spheres. If one of the spheres were enlarged to the size of the earth, its highest mountain would be only twelve feet high. First results were announced in early 2007, and they again confirmed general relativity.

CURVED SPACE-TIME: A STAGE SHAKING UNDER THE WEIGHT OF THE ACTORS

> Ay!
> There are times when the great universe
> Like cloth in some unskilful dyer's vat
> Shrivels into a hand's-breadth, and perchance
> That time is now! Well! Let that time be now.
>
> **—OSCAR WILDE,** *A Florentine Tragedy*

In general relativity, the form of space-time is determined by the matter it contains. Here the gravitational force finds its very origin, intimately connected with the structure of space and time in a way not realized for any of the other known forces in physics. Mathematically, all this is described by means of a curved space-time, a space-time whose degree of transformation between space and time, in the relativistic sense, depends not only on an observer's motion but also on the position in space and time. By this dependence on the observer's position, the concepts of special relativity are generalized. The theory is no longer constrained to what (inertial) observers moving at different but constant velocities along straight lines see. This assumption was a simplification employed in special relativity to understand the effects of different velocities, but it is not realistic: When we make observations, we are moving along complicated trajectories in space and time. We may be standing more or less still in a lab, but we are standing on the earth, which is rotating and orbiting the sun. The sun is moving, too, and so is the Milky Way. A general theory must be able to describe observers moving along arbitrary curves, possibly accelerating when forces are acting on them. General relativity does so by allowing for position-dependent transformations of space and time, thereby endowing space-time with a curved form.

The prime example of a curved space is the two-dimensional surface of a ball, or a sphere. It is a curved surface, and also closed in on itself, although the latter property is not shared by all curved spaces. What is illustrated by the sphere is the fact that lines on its surface must be curved, as seen from the surrounding space, in order to stay on the surface. Every straight line in space starting on its surface would immediately leave it. This behavior can be seen as a general consequence of curvature, even though abstract curved spaces do not need a surrounding space such as the three-dimensional one around the sphere. Space-time itself, for instance, is four-dimensional and would require an even higher dimensional ambient space. All consequences of curvature can mathematically be described without referring to such surrounding spaces—a convenient fact crucially exploited by Einstein in his formulation of general relativity. The relevant branch of mathematics, *differential geometry,* was founded on work by Bernhard Riemann in the nineteenth century.

Returning to the example of the sphere in its ambient three-dimensional space, we can see a further important consequence of curvature. When we move and change our position on a sphere, as we regularly do on the earth's surface, we are, as seen from the ambient space, forced to rotate. We usually do not notice this because, for one thing, Earth is very large, and moreover, we can rarely take this view from outer space. But the forced rotation can easily be visualized on a globe: The head of a person in Europe points in a different direction in space than does that of a person in America, even if both people are standing up ramrod-straight.[6] No such rotation would occur if one were moving on a planar surface such as the level floor of a room; it must be a consequence of curvature.

Space-time is curved by the matter it contains and should show effects comparable to those on a sphere. This is more difficult to visualize, for we now have a four-dimensional situation involving time as well. Our earlier analogy shows the most important consequence, which is directly related to the gravitational force: From special relativity we know that the transformability of space and time is connected with changes of velocities, just as the transformability of the three spatial dimensions is related to changes of angles. Just as a curved surface in space enforces a rotation when moving, moving in curved space-time should imply a change of velocities. Changes of velocities, or accelerations, are always caused by forces. The curvature of space-time thus implies the action of a force, which according to general relativity is nothing but gravity.

With this astonishing trick, Einstein was able to extend Newton's theory and erase its flaw. In Einstein's gravity there are no spooky interactions between distant objects in a direct way. By incorporating space and time—not as a rigid and given stage as Newton had assumed, but as a changing object with an inherent structure subject to physical laws—this action at a distance is forever banned from physics. Masses cause their directly surrounding space-time region to curve, upon which other masses experience a gravitational force as a result of the curvature. That this does not constitute action at a distance can be seen when the first mass is moving, causing the gravitational force on other masses to change. As shown by general relativity, this change does not occur instantaneously: The change in curvature has to propagate sufficiently far in space-time before it can reach distant masses. Physical interactions happen only locally, and what haunted Newton is resolved; a consistent theoretical underpinning has been gained. But perhaps the most impressive consequence of curved space-time can be experienced in cosmology, in which general relativity determines the temporal evolution of the universe itself.

LIMITS OF SPACE AND TIME: THE END OF A THEORY

Beware of asking for more time: no ill fate ever grants it.

—MIRABEAU

Promoting space-time from a mere stage, serving only to support the change of matter, to a physical object in the theory of relativity is a revolution (figure 2). The complicated interplay—matter curves space, and its own motion is influenced by curvature—leads to a mathematical description of unprecedented difficulty, keeping not only physicists but also mathematicians busy up to the present day. Fortunately, the theory is now sufficiently well understood to reveal many fundamental implications for our understanding of physical behavior, in particular that of the universe. The role of space-time, now seen as a physical object, is often compared to a novel in which one of the characters is the book itself. Consequences of such a novel would surely be surprising, though hard to imagine. Independently of imagination, consequences of the physical role of space-time can reliably be computed by means of the underlying mathematics. As we will see, this has even more ominous consequences in general relativity than are suggested by the example of the novel.

Before entering this dark chapter of so-called singularities, we will once more have a look at the relationship between gravity and the transformability of space and time. The division between space and

2. Objects move along trajectories in space-time, but space-time itself is changing. (*Orbita* [*Orbit*]), 2007. Sculpture and photograph: Gianni Caravaggio.)

time is influenced not only by velocity, but also gravity as caused by matter. For instance, time proceeds faster at higher altitudes, where the distance to the center of the earth is larger.[7] As before, such changes are usually imperceptibly small, but they do by now have technological relevance. The most precise atomic clocks experience a significant change in their rate when they are raised by just ten meters! As already mentioned, these effects, in addition to those due to velocities, must be taken into account if relativity is to be tested by means of atomic clocks on airplanes. (In this case, the gravitational effect is even larger than the velocity effect, and both are opposite to each other: At high velocities, clocks should be slower, but in an actual experiment with the typical altitudes and velocities of airplanes they are faster as a result of gravity.)

The global positioning system (GPS) is an example of an applied technology with crucial general relativistic effects. This is a system of twenty-four satellites, all carrying atomic clocks. They are distributed around the earth so that every place is almost always in reach of at least four satellites above the horizon. Each satellite sends out regular signals encoding its position and the time its clock measures. Comparing the signals of several satellites at a given point on the earth's surface allows one to compute one's position very precisely, usually with an accuracy of five to ten centimeters![8]

If one were to ignore relativity while computing the travel time of GPS signals, the resulting position measurements would be useless. After just two minutes of uncorrected clock time in orbit, clear deviations would be seen in the measured positions on Earth; waiting a single day would make measurements deviate from the correct values by up to ten kilometers. The role of general relativity in this system is of enormous importance, but it is very complicated and for a long time remained incompletely understood. The first GPS satellites were launched in 1978, but in 1979, 1985, and finally 1995—the year of the system's official inauguration—entire conferences were still being organized with the aim of understanding the role of general relativity in the GPS. Even so, an erroneous technical report was apparently published in 1996. Examples of difficulties are the role of synchronization and of the comparison of clocks on different satellites, and even the possible influence of the sun's gravity, whose strength would differ slightly on the day and night sides of Earth with their different distances from the sun. The latter turned out not to be important, given the current precision of clocks. (Which, to be sure, are very precise: They are rubidium clocks, whose measured times after ten days would differ by just half a nanosecond, that is, by half a billionth of a second.) But the changing gravitational field around Earth, and in particular its deviations from a perfect spherical shape caused by the oblong nature of the rotating planet, are of great importance to the clocks' precision. One can also ask whether special relativistic or general relativistic effects are dominant. At the altitude of GPS, about 27,500 kilometers, the clear winner is general relativity.

GPS applications are becoming more and more numerous, for instance with a cell phone feature especially popular in predominantly Islamic countries that always shows the precise direction to Mecca. The military applications for which the system was initially developed are by now clearly a minority, and those of geological exploration dominate, such as releasing light GPS devices in tropical storms to measure their temperature and pressure with position resolution. Also, the motion of the earth's crust as a consequence of tectonics can be followed precisely, and these data can perhaps one day be used to predict earthquakes. Even the motion and deformation of buildings or bridges under weight can be registered with the enormous precision of GPS. In agriculture, GPS is sometimes used to distribute fertilizer and pesticides very precisely, and archaeologists use it to find and map ancient historical sites. GPS, a child of relativity, now provides feedback on the exploration of general relativity itself: It serves to define a worldwide clock standard, which can then be used, for instance, to precisely measure the orbits in Hulse and Taylor's double pulsar.

All this shows the technological relevance of relativity theory, whose implications may be tiny in quantitative terms but are important given the current sensitivity of applications. When considering cosmic scales, on the other hand, the consequences of relativity become immense. Time is no longer rigid, as it was for Newton, but is influenced by matter in the universe. In extreme situations, this can, according to general relativity, imply that time itself comes to an end. The influence of matter on space-time is then so strong that time stops or space reaches an unsurpassable limit. Following relativity, this is supposed to have happened at the big bang (when we consider the universe in its backward evolution) or in black holes, where gravitational forces become so large that spatial or timelike distances shrink ever more and eventually vanish completely. Without timelike separations between events, time itself must die, and with it everything that happens. This dreadful conclusion applies to all material bodies as it does to the universe itself: Nothing can reach beyond such a point, a singularity.

What exactly would happen at a singularity? To see this, one must study the mathematical equations of general relativity, for they describe the structure of space and time. According to the continuous form of space-time, represented mathematically by differential geometry, those are *differential equations.* That is, Einstein's equations determine how the form of space-time near a point must behave on account of the matter present, or more precisely, on account of its energy density and pressure. Thus the equations correspond to a continuum picture, visualizing space-time as a curved and wrinkled sheet, albeit in four dimensions. In contrast to a real fabric woven from threads with gaps in between, the sheet of general relativity has no structure whatsoever when it is viewed on the smallest distance scales: hence the use of differential equations, which describe the change of space-time under the smallest shifts of position in it. These smallest changes are not atomic, but smaller than any possible size: They arise in a mathematical limiting procedure to conceptualize the continuum picture.

A differential equation can be visualized by its velocity field, in which an arrow is planted at every point to indicate the direction and size of the rate of change. In general relativity, the ground of the field in the simplest cases is a plane on which each point, by its two coordinates, determines time and the expansion of the universe at that time. A solution to the differential equation is a curve in the plane required to follow the direction of the velocity arrow at each point. One can often represent the solutions graphically, as in figure 3, but there are also mathematical methods to directly represent them in the form of a function, or numerical techniques to find such curves with computers.

The graphic construction shows that the velocity field is not sufficient to determine a unique solution; it indicates, after all, just the direction to move at each point. First one must know where to start: A point, the initial condition, must also be chosen. If this is done, one obtains a unique solution in most cases, but the form of initial conditions can, depending on the problem analyzed, take more complex forms than a single point.

In cosmology, the velocity field of cosmic expansion is deter-

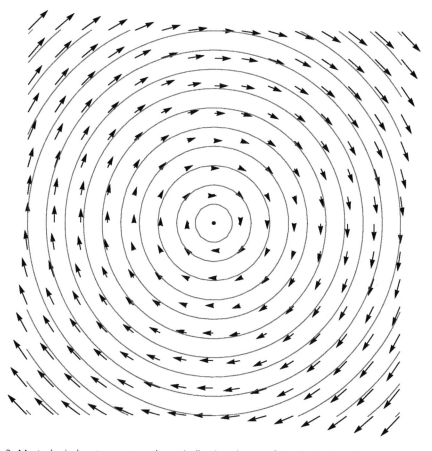

3. Most physical systems are mathematically given by specifying their rates of change every-where, formulated as a differential equation. The rate of change can graphically be represented by arrows such that a solution to the problem is a curve always taking the direction of the arrows. In this example, the velocity field of a vortex is shown, whose solutions are circles around the center point.

mined by matter in the universe. Small changes of space-time caused by the smallest shifts in position are then given by the size of the energy density and pressure of the surrounding matter. As the initial condition we can posit a state resembling what we see now in the universe, a mathematical configuration with expanding space and diluting matter content. This leaves different possibilities for future development: If the amount of matter is small, below a certain criti-cal density, the universe will expand forever and become more and more diluted. But amounts of matter above the critical density would exert a stronger gravitational force, which, in the distant future, can first stop the expansion of the whole universe and then make it collapse on itself. In this case, there will be a time when the universe turns around, like a rock thrown up in the air, and then contracts. To decide which case will be realized for our universe, we must determine the exact amount of matter—not an easy task. We can estimate and add the masses of all stars and of the hydrogen gas in and between the galaxies, but, as we will see later in the context of observational cosmology, there are further forms of matter and energy more difficult to quantify. Their origin is not known, but observations indicate that the total energy density is very near the critical one. It thus remains uncertain whether our universe will expand forever or one day collapse.

The future is always uncertain; we should therefore have a look at what happened in the past. Owing to its expansion, the younger universe was smaller than it is now, and very long ago it was hotter than any kind of matter existing on Earth or even in stars. Under such extreme conditions, the behavior of matter is little known, but general consequences for the universe only weakly depend on it: A detailed analysis of Einstein's equations shows that some of the spatial or timelike distances must be unimaginably small at the moment of the big bang (and in black holes). One can see this in the velocity field of figure 4, which pushes every solution curve to a point of vanishing volume. The energy density of matter—the ratio of the total energy to the occupied volume—becomes infinite: One divides by zero when an extension, and thus the volume as a product of extensions, vanishes. Here we have the physical dilemma that no matter can persist at infinite density; also, such a divergence is mathematically so problematic that it leads to the breakdown of Einstein's equations. Even if we tried to ignore the problem of infinite density, the equations would tell us that the form of space-time changes by an infinite amount even at the tiniest change of position (or time). This cannot possibly be a useful coherent structure: Space-time is torn asunder at the singularity.

Singularities constitute a serious problem, a theoretical misdemeanor that will eventually force us to abandon general relativity as a fundamental description, or to extend it. The tremendous consequence that space and time reach an end in their theoretical description does not mean a physically predicted beginning or end of the world. Even though the mathematical equations show that a point is reached at which all distances vanish, the equations themselves then

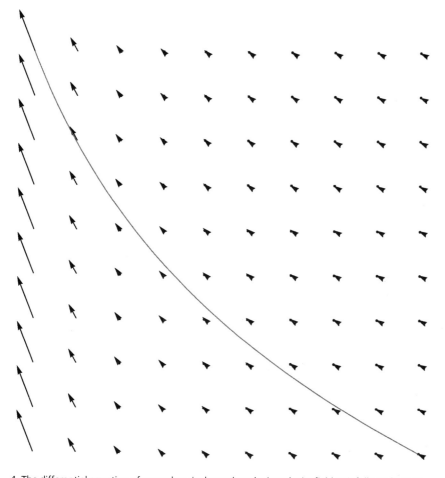

4. The differential equation of cosmology is shown here by its velocity field as it follows from Einstein's equation. The volume of the universe changes horizontally; vertically, the density of the matter content grows. The closer one comes to the left border, where the universe volume vanishes, the longer the arrows become. Every solution curve along the arrows is inescapably drawn to the border. Moreover, the matter density grows without bounds and becomes infinite for vanishing volume at the left border.

lose all their meaning. The theory becomes unreliable at this point and simply can no longer be used for predictions; it leads us to this singular place near the abyss, but then leaves us alone with the question of its meaning and of what lies beyond. From then on, we have to look for a new guide.

Just like Newton's theory, Einstein's gravitational law is flawed, but much more seriously so. While Newton was uneasy with the animalistic tendencies of his theory, the shadow looming over general relativity is decidedly more ominous. In the words of the physicist John Wheeler, general relativity contains the seed of its own destruction—which is a condition for greatness: "All great things perish on their own, by an act of self-elimination."[9] Indeed, general relativity is unique not just in its power but also in telling us its own limits in a strong, undeniable way. Singularity problems of this form exist nowhere else, since only general relativity, so far, addresses the behavior of space and time without bias. By being such a strong barrier, the singularity problem attracts much interest in trying to overcome it by more complete theories. In the comparison with a novel, this means that the book itself is not only a character but will, in the course of the action, perish.[10]

Suicidal theories are very uncommon in physics, and so it is not surprising that consequences are often misinterpreted.[11] Even Einstein thought that limits of space and time due to singularities appear only in special cases and not in general situations. In his time, such an opinion was not entirely unrealistic, for not much was known about the various mathematical solutions of general relativity. (Einstein did, however, make fruitless attempts to prove the irrelevance of singularities.) Only after later studies by Stephen Hawking, Roger Penrose, and others in the 1960s did it become clear that limits of space-time cannot be avoided within general relativity and must be taken seriously.[12] Mathematical solutions, required only to be compatible with the present form of the universe, have at least one singularity—a limit to space and time—where general relativity itself loses its validity.

LACK OF REPULSION: THE DANGERS OF BEING TOO ATTRACTIVE

> We saw this for the simplest of all natural phenomena,
> gravity, which does not stop striving and pushing toward
> an extensionless center point, which when reached would
> be its own and matter's demise, which does not stop even
> if the whole world would already be conglomerated.
>
> —ARTHUR SCHOPENHAUER, *The World as Will and Representation*

Singular as the phenomenon of gravitational collapse may be, it has a physical origin. As in Newton's theory, the force in Einstein's law of gravity is always attractive and we need other kinds of counteracting forces to bring about stable situations. (There is a possibility of repulsion in general relativity as a result of strongly negative pressure, as we will see later. But these forces cannot be large enough to completely compensate the attraction and prevent singularities.) To swim or fly at constant height, we need buoyancy forces. In order for us to stand stably on Earth and withstand the attraction toward the center, gravity must be balanced by forces of the solid ground. Earth itself does not collapse on itself because it consists of solid matter (or liquid at its core). If Earth were heavier, its internal pressure and temperature would rise and cause partial melting and evaporation, as happens on gaseous planets such as Jupiter and Saturn. Sufficiently strong repulsive forces between the gas atoms exist to counteract gravity at high pressure. Were those giant planets to become even more massive, the gravitational pressure would increase. As they became more and more compressed, the increasing density would at some point bring pairs of hydrogen atoms close enough for them to fuse into helium. The planets would then become stars, which by the fusion process produce enough energy not only to shine but also to help compensate gravity by internal heat.

At even higher masses the heat pressure will no longer suffice. At first, there is a force of quantum mechanical origin to help out matter, as in white dwarfs. This force arises because one cannot compress a gas of electrons beyond a certain point of density, even though the electric repulsion would be neutralized by the positively charged protons in the star matter. Alas, this force finds its limits if the total mass is increased further, for electrons begin to react with the protons of nuclei at high pressure. The gravitational pressure keeps increasing, and from some point on, atoms will be so compressed that protons in the nucleus will fuse with orbital electrons to form neutrons. A neutron star is the result, with a core of pure neutrons. Such matter is extremely dense because the safe distances between the nuclei and electrons of usual matter have been trespassed; it can exist only under the severe conditions of a strong gravitational pressure, for otherwise most of its neutrons would soon decay back into protons and electrons. More details about these exotic stars can be found in the discussion of black holes in chapter 6.

But not even this hardy state can prevail against gravitational pressure when mass is increased further. From then on, no known force could counter the rise of gravity in this cold war of forces. Equilibrium is no more; all chunks of matter do nothing but attract each other, and they collapse to a black hole, another manifestation of the singularity. In its interior, the unrestrained collapse of all infalling matter generates ever higher densities and temperatures, and then at last—through matter's influence on space and time—the end of time itself, once all the matter of the former star has collapsed. "What an ending! What an appalling ending!"[13]

Here we have the physical reason for the existence of singularities in general relativity: the attractive nature of gravity without powerful repulsive forces. That bold leap of relativity beyond Newtonian physics, fueled by the changeable structure of space and time, turns out to be a very risky one. Space-time becomes a dynamic object, subject to mathematical equations. Unfortunately, only in exceptional cases do these equations have solutions defining the universe at all times. Most solutions, including realistic ones, lose their validity at some point, so space-time as described by them comes to its own end, too. Such singularities exist because general relativity, while eliminating Newton's action at a distance, does not change the purely attractive nature of the gravitational law. Moreover, matter attraction is allowed to influence even space and time by the daring innovation of general relativity. It rather fiendishly unleashes gravity on space and time, which now must tolerate the whims of matter without any safeguard by a counteracting force to halt the full collapse. To solve this not only aesthetic but also decidedly fundamental problem, we must extend the theory once more, and take a leap beyond.

There are comparable cases in which a theory leads to singularities, which can be eliminated by suitable extensions. Liquids such as water, for instance, can often be described quite precisely by assuming a continuous distribution, ignoring the composition of molecules. One might use this approximation for the flow through a pipe. But for small flow rates, and if the pipe ends in a faucet, this description breaks down: The continuous distribution decays to single droplets at the end of the faucet. From the viewpoint of the continuum description, the breaking apart appears as a singularity, for some quantities such as the surface tension would diverge at the time of separation; continuum equations then lose their meaning. In this case we know what causes the problem: When a droplet splits off, nothing really violent is happening. We merely have to consider the molecular nature of matter; and if the gravitational tension surpasses the cohesion of water molecules, water can split into separate drops. One has to extend the continuum theory by a more fundamental theory taking into account the molecular nature of matter. Such a theory is more realistic but also mathematically more involved. Its key advantage: It does not lead to singularities in its solutions and can describe, for example, the separation into droplets.

There are significant differences between gravity and this water analogy, in particular due to the more intimate involvement of space and time in general relativity. But this analogy conveys an important lesson: There is no avoiding an extension of general relativity. Singularities show the limits of the theory, and to understand their role one must find an encompassing theory that regularly describes the would-be singularities.

Only then can we check whether limits such as those in black holes or at the big bang are truly physical limits of space and time, or merely limitations in our theoretical description. Here we have a situation similar to Einstein's before he formulated general relativity: We do have a very successful theory of gravitation that is not in conflict with any observation. And yet we know that this theory cannot be complete; we must find an extension based on general principles. Apparently, we are not as lucky as Einstein, who—an intense but short decade after perceiving the problem that Newton's law is incompatible with the transformability of space and time—was able to propose general relativity as the solution. Special forms of singularities were found in solutions just after 1915, for instance in the solution for a nonrotating black hole determined by Karl Schwarzschild in 1916. The general existence of singularities in solutions to general relativity was demonstrated in the 1960s. In spite of numerous more recent efforts by far more theorists than in the years between 1905 and 1915, no complete theory has yet been formulated, a theory at same time preserving the successes of general relativity and solving its singularity problem.

There is one belief guiding the majority of researchers: Next to general relativity, quantum theory is the second pillar on which twentieth-century physics is built. Quantum physics describes objects in extended, wavelike forms rather than pointlike particles, and establishes as a consequence the atomic, discrete structure of matter. Quantum theory is indispensable for a correct description of matter and implies impressive phenomena. In astrophysics, as already mentioned, the quantum nature of matter leads to new stable states such as white dwarfs and neutron stars that rely on repulsive forces caused by quantum physics. Sometimes these forces suffice to compensate the attraction of gravity, but this is not the case for large masses. What is completely left out of this picture is the quantum nature of gravity and space and time themselves, in addition to that of matter. Had we ignored the quantum theory of matter, even white dwarfs and neutron stars could not exist stably, for no repulsive classical forces would occur at their densities. This would be in conflict with observations that have confirmed the existence of such objects in the cosmos.

The quantum physics of gravity is ignored in all these investigations, since such a theory is not yet available. All the known but incomplete approaches to its formulation, moreover, are mathematically so complex that they cannot yet be applied to objects such as neutron stars or black holes as they exist in space. But there are situations in cosmology where strong evidence of new counterforces exists, forces arising in a quantum theory of gravity and possibly leading to a theory free of singularities. An understanding of far earlier times before the big bang, as well as of the interior of black holes, then comes within reach. But before we can turn to these developments of recent research, we will need the basic concepts of quantum theory.

2. QUANTUM THEORY
UNCERTAIN STABILITY

For as its belt sparkled and glittered now in one part and now in another, and what was light one instant, at another time was dark, so the figure itself fluctuated in its distinctness: being now a thing with one arm, now with one leg, now with twenty legs, now a pair of legs without a head, now a head without a body: of which dissolving parts, no outline would be visible in the dense gloom wherein they melted away. And in the very wonder of this, it would be itself again; distinct and clear as ever.

—CHARLES DICKENS, *A Christmas Carol*

Quantum theory is a universal framework applying to all systems, even macroscopic ones. It is indispensable for a correct description of atoms, the realm of its initial discovery. Its importance has been verified experimentally many times, and as with general relativity, no deviations between observations and its predictions have occurred to date. And yet these two dominating theories do not provide a complete picture of physics, for in their current formulations they are incompatible with each other. This is of particular importance in extreme situations, such as the big bang or black holes, in which a quantum theory of gravity seems to be the last hope to cope with singularities.

In addition, one has to fight with the peculiarities of quantum theory: for instance, the abrupt change of the state of a system even as it is being measured, and the impossibility of complete knowledge as a result of fundamental uncertainty. The wave function, describing all the information contained in a quantum system, must brutally collapse during a measurement, condemned by the uncertainty principle to an incomplete confession. But it is worthwhile to accept this challenge, for quantum theory does have welcome consequences, too—in particular, concerning singularities.

ATOMIC STABILITY:
THE PRICE OF LIVING LONG

Conceptually, if not historically, the beginning of quantum theory is marked by a stability problem not unlike that of gravitational situations. A hydrogen atom consists of a positively charged proton as the nucleus and a negatively charged electron in orbit. An electron's mass is only about half a thousandth as much as the proton, suggesting the classical picture of a light electron orbiting around a heavy proton. By electric attraction, the electron is forced to move in a circular manner, just as gravity compels the moon to circle the earth. The radius of the electron's orbit may be seen as the radius of a hydrogen atom, and has a size of about 50-billionth of a millimeter.

This picture has a serious problem. According to Maxwell's theory of electromagnetism, an electric charge moving along a trajectory that is not a straight line must radiate waves. As a general phenomenon, this is similar to the emission of gravitational waves in a double pulsar, whose radius shrinks in time as a result of the energy loss; as described in the preceding chapter, measuring consequences of the energy loss has allowed impressive tests of general relativity. In the case of electromagnetic radiation by moving charges such as the electron, the analogous phenomenon is much more common; it is often exploited, for instance when generating X-rays or cell phone signals.

The emission of waves also means that the electron loses energy through its motion, can no longer resist the electric attraction, and thus approaches the proton. Energy lost is sent out by the atom as visible light or in the form of invisible electromagnetic waves. Maxwell's equations can be used to compute the amount of energy loss, showing that the emission must proceed much faster than that of gravitational waves by a double pulsar: A bound electron would lose all its energy after just a fraction of a second and then collide with the proton. In contrast to the slow shrinking of a double pulsar, this behavior is in disastrous conflict with the observation that hydrogen atoms exist stably for extremely long times. The classical picture fails miserably—not only for hydrogen but for all atoms. Matter, if it obeyed classical laws, could not be stable, because its constituent atoms could not exist even for a second. This problem may not be as damning as that of singularities in general relativity, where even space and time would reach their limits and literally nothing could survive. On the other hand, the problem of atomic stability is more concrete; we would, after all, be affected by instabilities of matter in a very personal way.

Quantum theory resolves the stability problem of atoms in an elegant, but unfortunately quite counterintuitive, manner. To see how this can happen, we first have to shed some light on the general principles of quantum theory.

THE WAVE FUNCTION: FLUCTUATIONS, SUPERPOSITIONS, LIMITATIONS

> Now that Yossarian looked back, it seemed that Nurse Cramer, rather than the talkative Texan, had murdered the soldier in white; if she had not read the thermometer and reported what she had found, the soldier in white might still be lying there alive exactly as he had been lying there all along, encased from head to toe in plaster and gauze. . . .
>
> **—JOSEPH HELLER,** *Catch-22*

Quantum theory implies that protons, electrons, and the rest are not particles of the classical kind, visualized as pointlike or as solid spheres. Instead, what we call a particle corresponds to a wave function: an extended object, spread-out and diffuse due to its wavelike character. It does not have a distinct surface as a ball does; outside a central region it slowly fades away. Just as a water wave approaches lazily without allowing one to say when exactly it has passed a swimmer, the wave functions of quantum theory do not have sharply defined boundaries. Even though their tails are usually of tiny height, they can extend to regions far away from the center, the place best seen being analogous to the "position" of a classical particle. This is the origin of many phenomena that at first sight may appear impossible but can in fact be observed.

What may also seem counterintuitive is the interrelation of different objects described by wave functions. While classical particles can be put at separate places like billiard balls, and hit at different speeds to watch their motion and collisions, a single wave function already occupies all of space. Even though it may have nearly vanishing height far from a center, it does not have to be exactly zero. When a second wave function is positioned anywhere in space, its background differs from what it would be in the absence of the first wave. This is the case even if the respective central regions are far apart, which would allow us to ignore all kinds of attractive or repulsive forces. Just the presence of the first wave function, anywhere in the universe, has a certain influence on another wave function.

It is truly bizarre that completely different states of a single object can be superposed and coexist. Erwin Schrödinger, one of the prime movers in the early days of quantum mechanics, has described this pointedly in a thought experiment now called "Schrödinger's Cat." A cat is locked away in a cell enclosed by nontransparent walls, also containing a potentially deadly mechanism: a radioactive substance whose radiation, via a detector, triggers a hammer to break a flask containing poisonous gas. A substance of weak radioactivity has been chosen, which will only rarely trigger the detector and thus the murderous device.

If and when radiation will trigger the detector is, like the whole atomic world, governed by quantum mechanics. Macroscopic objects such as the cat must, in the end, obey the same laws of quantum mechanics, but this does not often make itself noticeable on the large scales we are used to in everyday life. That is why classical physics was able to describe everything so well until precise measurements revealed the true finesses of the quantum world. But here, via the hammer and the poison, the quantum moods of the radioactive substance are transferred directly to the macroscopic world of the cat, now helplessly delivered to the laws of quantum mechanics. There is no law for the exact time of the substance's decay; there is a law only for the decay probability. According to quantum mechanics, this means that the atoms of the substance are not in one fixed state, either already decayed or not yet so, but in both of them! The atoms are in a superposition of both possibilities, and the poor cat, too, is hanging on in a zombic combination of life and death. When and only when a measurement is performed, for instance by peeking in to check on the fate of the cat, is a definitive state realized where the cat may or may not succumb. In the language of quantum mechanics, the wave function then collapses from its superposition into a state with a fixed measurement outcome.

The end result, however, is undetermined. Once the wave function has collapsed, to be sure, a second check would not lead to a different state; if the cat is found dead, it cannot be revived simply by looking away. After the measurement, the result is definitive, but before the first measurement is done, there can be only likelihood statements about the result. Performing the experiment with a hundred cats, all in the hands of their own murderous device, quantum mechanics would tell us how often we find living cats and how often we find dead ones after a certain amount of time; but it would not tell us which of them would be alive and which ones dead. It is not even possible to gain additional information before observing the cats. By many tests, quantum mechanics has been shown to be a fundamental theory. Its probabilistic nature, telling us only likelihoods of measurement outcomes, is not a consequence of limitations of the theory; it cannot be avoided in principle. The more varied the incompletely determined possibilities, in contrast to just a single determined one as would be realized in classical physics, the larger the role of quantum theory. One then speaks of strong quantum fluctuations, meaning that results can differ strongly from one measurement to the next.

In the classical world, we are not used to superpositions. To be sure, there are sometimes doubts before a suspicion becomes brutal certainty, but we are convinced that some possibility is realized; we simply do not know which one. In quantum mechanics all this is different. Here, too, there is uncertainty, but one can prove mathematically that superpositions bring in a new quality. There is no doubt that microscopic objects really do exist in superpositions of different states; this has been confirmed by many measurements. Superpositions have not been observed for macroscopic objects such as a cat, even though it would in principle be possible. Bringing and keeping larger objects in a superposition is simply more complicated. This explains why we are not used to this feature from our common classical experience, and thus our intuition struggles with the concept of superpositions. The following passage presents a further visualization of a superposition together with its eventual collapse:

She was a little spider. From her own thread she had made a web, and she was hungry. Besides, what else could she be doing? Now she was lying motionlessly in her corner, her legs stretched out and all her eyes staring up through the web, waiting for something to happen. She did not think much, for, to be honest, she did not even know what a spider should be thinking about. She was just passing her time, in expectation of this unmistakable event announcing that her hunger would be stilled.

Before long, something crossed her mind. Not a thought, for she still did not know what a spider could be thinking about. From somewhere, she did not know how, she remembered that spiders were sometimes supposed to eat their own, females the males. She shuddered. At least she would be at the winning end. She had grown hungrier, as she realized.

She did not know how long she had been lying there, still in her comfortable position. In fact, her feeling for time was not so well developed. Although she instinctively knew that things changed—her stomach was often empty as the web, but every once in a while there was a feast—she had no concept of time. Was it more boring to lie waiting like this, or when time would not exist at all? What a crazy idea. She almost fell asleep.

At last, her full attention was called upon. She had heard a shrill sound which, as she fancied, had made her web vibrate. This noise, so familiar and yet alien to her little world, had only briefly made her anger boil over about the disruption of her peaceful rest. Then, knowing what this meant, she jumped up in excitement: soon, she would no longer be hungry. Although she did not know how, she knew the meaning of the sound, and clearly made out a few distinctly separated words: "Dinner's ready!" The little spider died, unremembered, and she ran downstairs.

Just as the apparent contradiction of the last sentence here resolves the superposition of "she" (a girl pretending to be a spider) and "the spider," a measurement at a quantum mechanical system renders its state definite by the collapse of the wave function.

Superpositions play an important role in quantum theory and its applications. They would be an integral part in the construction of quantum computers in which classical bits, the elementary carriers of information, are replaced by quantum bits—"qubits" for short. As quantum superpositions of states, they would allow more combinations than classical bits. If a stable construction can be accomplished, keeping superpositions undisturbed for a sufficient amount of time, calculations could be performed much faster than with traditional computers. Different parts of the wave function could, in a sense, participate simultaneously in a calculation instead of having to process a long line of consecutive contributions as in a classical computer. This capability can lead to entirely new computational algorithms.

Besides the possibility of superpositions, quantum theory shows several important differences from the principles taken for granted in prequantum, classical physics. In contrast to a classical particle, objects described by a wave function cannot be assigned distinct positions or velocities. Measurements of position or velocity are always subject to an inherent uncertainty that cannot be eliminated by improving the measurement apparatus. In a single position measurement, one could, in principle, find any value within the region where the wave function has nonzero height. Performing many measurements, one will find a different value in that region each time; a large number of measurements leads to a distribution of results in which the frequency of a specific position measured is given by the height (or rather, its norm squared) of the wave function at this place. Certainly, one would not expect to localize a wave function far away from its center except in rare cases; but since the height there can still be nonzero, one may sometimes take a place far from the center as the actual position of the wave. In contrast to a water wave, which, by daylight, can be seen in its full extension, measurements of a quantum theoretical wave function are random tests. One checks out the wave function just a few times as if trying to feel its shape at night. And before one can measure the wave often enough to find its precise position, it will have moved away or—just by having been touched—changed its form.

The height of a wave function at some point, so says quantum theory, determines how often one would take this place as the actual position of a particle. In mathematical terms, this distribution is a probability: the likelihood of the result "position x" is given by the height of the wave function there. If this height at some place is larger than it is somewhere else, attempts to feel the wave would more often make one take this place for the position of the particle and one is more likely to identify this place as the actual position. Similarly, probabilities can be found for measurements of velocity or any other quantity of interest. One would think that one could then simply determine the whole wave function by measuring the position sufficiently often, deriving from the individual results for all points in space the shape of the wave function. When each position measurement is plotted in a diagram, darker regions would be seen where the particle was found more often, surrounded by lighter shades where the particle would enter only rarely. Indistinct but noticeable contours would result that one could interpret as the shape of an object measured.

But once again this is too classical a picture of a wave to be applicable in quantum theory. Such measurements would be so sensitive that their process itself, as a physical procedure, must influence the state described by a wave function. After, say, the position has been measured, a wave function is different than before. With most measurements, this goes so far that the wave function is said to "collapse" due to the measurement process—similarly to Gawain's "quantum bed," the *Lit marveile*:

> He [Gawain] entered. Its pavement shone smooth and clear as glass. Upon it stood that fabulous Bed: Lit marveile! . . . He went at a peradventure. And as often as he made a step, the Bed moved from where it was. . . . "How shall I get at you?" he wondered. "Are you set at dodging away from me? I shall teach you, if I can pounce on top of you!" At that moment the Bed stood still in front of him. He took a flying jump and landed plump in the middle. No one will ever hear again of the speed at which that Bed went crashing from side to side!
>
> **—WOLFRAM VON ESCHENBACH,** *Parsifal*

We are thus prevented from ever determining the complete wave function by many repeated measurements of an object. Very delicate methods that handle the wave function more gently do exist, but no measurement is possible that exerts no influence whatsoever. As a principle problem it cannot be evaded by improved technologies; every measurement must interact with the measured object if it is to provide any information at all.

In everyday life, we can investigate the shape and texture of surfaces by physical touch, leaving sturdy objects unharmed. But sensitive objects are changed or even destroyed by contact and cannot reliably be examined in this way. One can use more delicate measurement tools instead of one's fingers, such as optical methods. But even light carries energy and must interact with the object in order to survey its properties in a useful way, exchanging energy. Even dim light is too energetic to leave the fragile wave functions of atomic physics unharmed. Every measurement changes its object, implying a fundamental limitation to information gain. What *can* be achieved by cleverly sophisticated technologies is a determination of properties of the wave function in as complete a way as possible—within the limits allowed by quantum mechanics. Such delicate measurements are called "quantum nondemolition" since they attempt not to destroy the wave function. But even in this case, it is impossible to determine the wave function completely. Strictly speaking, it does not represent a physical object, fully observable in all its details; it is rather a mathematical description of all the accessible properties.

The influence of a measurement on an object has a further well-known and important consequence: Werner Heisenberg's uncertainty relation. Uncertainty relations occur whenever measurements of different, so-called complementary quantities are undertaken. The position and velocity of a particle as described by a wave function provide the best-known example. In this case, the uncertainty principle can intuitively be demonstrated by an explicit, if conceptual, measurement process. A microscope, for instance, allows one to measure the position by scattering light off a particle. Without the particle, light would simply move straight from the source to a detector; when the light is scattered off a particle in different directions, the particle can be located. This very scattering, essential for a successful detection, is an interaction influencing the object scrutinized. Light carries energy, and the scattering process changes the particle's motion. If the particle was initially at rest, say, it will move slightly after some light has ricocheted off it. For the macroscopic objects we usually deal with in life, this change is imperceptibly small, but it plays a crucial role for a microscopic object such as an electron.

During a very precise measurement of position, the velocity of a particle must inevitably change; these two quantities are exactly the complementary ones of this example. There is an effect on the velocity, the more precisely one attempts to measure the position: Higher resolution requires larger energies of the light used, as becomes clear when replacing light with electron microscopy. More energetic light affects the velocity more strongly; measuring the velocity becomes less precise for a more precise position measurement. Here we see the uncertainty relation for complementary quantities: High precision in the measurement of one of them must be paid for by reducing the precision for the complementary one.

The uncertainty relation in quantum theory is a law of nature, and it cannot be avoided if quantum theory is valid.[1] Despite these possibly unwelcome measurement limitations, the wave functions of quantum mechanics have important consequences for atoms. An electron's wave function, in contrast to the classical understanding of a particle orbiting around a nucleus, results in a limited set of possible energy states: Only very specific energies—infinitely many, to be sure, but only in a discrete set of isolated values—can be realized.

ATOMS: STABILITY FROM UNSTEADINESS

An electron of a given energy, bound in an atom by electric attraction, has a spherical wave function of fixed wavelength surrounding the nucleus. How quantum mechanics leads to the realization of only a discrete set of energies can be seen in the simplified context of a wave function on a circle surrounding a central point. The wavelength, the distance between neighboring hills of the wave, is related by quantum mechanics to the electron's velocity or momentum: The faster the electron, the more rapidly the wave changes, and so the wavelength is shorter. If we imagine following such a wave once around the center, laying wave crests next to wave troughs at the required distance, the form will generally change when we return to the starting point after a single orbit. A peak no longer falls on its initial position, unless an integer number of wavelength intervals can be fitted into the circumference of the circle (see figure 5). If there is a mismatch, it causes a spatial shift of the new crests, repeated every time we orbit around. At some point, a new crest falls on an old trough, and the negative heights in the trough exactly cancel out the positive heights in the crest: The whole wave disappears in destructive interference. Only in exceptional cases, for those finely tuned circles whose circumference is an exact multiple of the wavelength, can the wave be in such harmony that each orbit will lay crests exactly on crests and no destruction will occur.

In this way, an electron on a circle of fixed radius can move only with velocities in a discrete set; otherwise its wave function, and thus the electron, could not exist. At this stage, we have only considered the wavelike nature of quantum mechanics, not properties of the forces confining the electron to the circle. Forces provide additional conditions, for they tell us how the electron is being accelerated or slowed down. For motion along a circle, moreover, a balance must

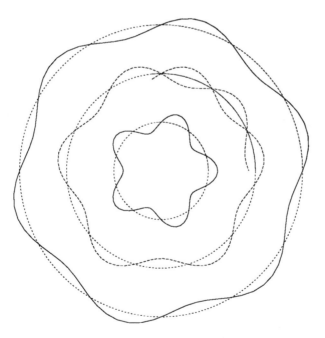

5. Drawing a wave of a given wavelength along a circle is possible only for a discrete set of specific radii. If the radius and wavelength do not match (as shown by the dashed wave), the wave after one orbit is displaced from its original position to the degree that its variations away from the circle begin to cancel each other out.

be realized between the force exerted by the center, attracting the electron, and the electron's normal tendency to follow a straight line, moving outward from the circle. A further relationship arises, telling us how fast the electron must move when it has a certain distance from the center. Now, there are two conditions for the electron, both tying the velocity to the radius: one relationship that also exists classically, since it refers only to the force, and the new observation in quantum mechanics, requiring the wave to fit wholly on the circle. Each condition takes a different form, and can thus be realized only for special radii. Changing an allowed radius slightly will render the wavelength out of tune with the circumference, and it is impossible to satisfy both conditions. In particular, the requirement of integer multiples of the wavelength to fit on a closed circle, a condition arising only due to the wavelike nature of quantum mechanics, implies that only discrete, isolated radii can be occupied.

For an electron orbiting the nucleus of an atom, these are the orbits around the nucleus the electron is "allowed" to take, for only then can a nonzero wave function result. Since the velocity has already been tied to the radius, and both determine the energy, energies must also lie in a discrete set, isolated from each other. While motion in an atom is more complex than circular motion, the same reasoning regarding waves applies: Electron energies in different orbits cannot be arbitrarily close because of the fixed wavelength separating the wave function's crests; changing the energy slightly causes the wave crests to go out of synch. If we want to change the energy, we must do so by a certain finite amount to ensure that crests fall on crests again. As an immediate consequence of the wavelike nature of electrons, this ladderlike form of allowed energies—the discreteness of the energy spectrum—distinguishes many properties of quantum theory (see figure 6). By contrast, in classical physics, which lacks the extra condition for the wavelength, one would expect that simply choosing the radius of the electron's orbit would allow one to provide any arbitrary energy value.

In particular, energies cannot be arbitrarily small, bringing us back to atomic stability. Classically, the electron, like any electric charge in spherical orbit, would emit all its energy and quickly fall into the nucleus. For a quantum mechanical, wavelike electron, this condition does not arise, owing to the restrictive nature of the additional condition that already gave rise to discrete energies. More intu-

↑Energy

6. For an electron bound in a hydrogen atom, there are infinitely many allowed energy values, which occupy a limited range. They must accumulate at some point, at which the discrete distribution looks like a continuum. Here, the special properties of quantum theory are suppressed—in stark contrast to the low energy values, where long distances between the allowed energy values are realized.

itively, we can relate the important contributions of quantum theory in this context to the necessity of counteracting forces. At the same time, this will lead to a further visualization of wave functions and their role.

To stabilize a hydrogen atom, with a single proton in its nucleus, we need a new force counteracting the classical energy loss from electromagnetic radiation emitted by the orbiting electron. Exactly such a force arises as a result of the spread of the wave function when the distance between its center and the proton in the nucleus is smaller than the width of the wave. While the main part of the wave function remains concentrated mainly to one side of the proton, a sizable fraction is located by the opposite side. (The proton itself is not a classical point particle either, and must also be described by a wave function. Its spread, however, is much smaller because of the larger mass of the proton and does not play a role here: The heavy proton does not whirr around as wildly as the light-footed electron, and thus has a much more precisely defined position.)

Since the wave function describes the probability for the electron's position, there is now a possibility for it to change place to the opposite side of the proton. There, it will still be electrically attracted toward the proton, but the force, as seen in figure 7, points

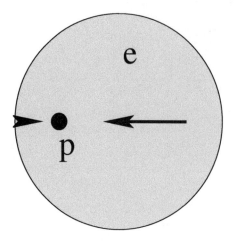

7. The extended wave function of an electron means that at any given time, it does not have to be strictly on one side of the proton. Electric attraction from different sides intuitively implies repulsive forces (opposing arrows) due to the quantum mechanical nature of particles, bringing about the stability of atoms.

in the opposite direction compared to what it was on the original side. Thus emerges a repulsive force, counteracting the classical electric attraction of two point particles to help stabilize atoms. Repulsion is largest for an electron wave function fitted around the proton like a sphere, a configuration that indeed corresponds to the quantum mechanically computed state of lowest possible energy.[2] Here we have an example of how quantum theory can lead to new forces, and consequently to stability.

In this way, separate states of atoms exist, with clearly distinguished energies not allowed to be arbitrarily small. Among those, the stable state, called the *ground state,* has the lowest energy; all others are more energetic and excited. From an excited state the atom can emit energy in the form of light, causing the electron to occupy a less energetic state. Thanks to the discreteness of possible energy values of an atom, emitted energies can take only certain sizes. One can directly observe this from properties of the emitted light—its frequency is proportional to the energy it carries away—and then compare them with calculations in quantum theory. This has provided many tests of quantum theory, so far all passed brilliantly.

Such data already existed before quantum theory was developed in the early twentieth century, thanks to the straightforward measurability of discrete emission and absorption spectra. Finding an explanation was a prime motivation for the first researchers of quantum theory. Excellent experimental data, unexplainable by classical physics, thus existed when quantum mechanics was developed, which is different from the situation for general relativity. Researchers found important clues as to how exactly classical physics had to change. Otherwise, quantum theory, with its strange consequences, would likely never have been found, let alone accepted.

THE CLASSICAL LIMIT AND EFFECTIVE FORCES:
TEARING DOWN WAVES

Extensions of already established theories are supposed to explain new observational data, but they must not give rise to disagreements with phenomena already explained by an older theory. Achieving this consistency is no simple task, for the tremendous progress of physics since Galileo Galilei has made existing theories very successful, explaining numerous phenomena that fill whole books and even libraries. It is impossible to reevaluate every single explanation in order to show explicitly that consistency remains realized. Instead, one can often cleverly make use of the fact that extensions of a theory usually introduce a new parameter: a new constant of nature. Phenomena explained only by the new theory allow one to determine the value of such a parameter by comparing theoretical predictions with measurements. The old theory, on the other hand, can be realized as a limit of the extended theory in which the new parameter takes a certain value that is fixed but incompatible with the newly explained observations; this incompatible value is usually zero, but it may also be infinite, which is actually allowed in such a limiting process.

For relativity, the new parameter is the speed of light, implicitly assumed to be infinitely large in Newtonian physics. Thus the unwelcome action at a distance in Newton's law of gravity: Without any upper limit for propagation velocities such as that of light, a mass can instantaneously act on other masses no matter how distant they are. In quantum theory, the new parameter is *Planck's constant,* which determines the size of discrete quantum jumps of energies. (Even before the development of quantum mechanics, this constant was introduced by Max Planck in the context of black body radiation, as discussed below.) Accordingly, this parameter can be measured in the emission spectra of atoms, reflecting the discreteness of excited states. By setting Planck's constant to zero, one would eliminate the distances between different energy levels, pushing them close to each other, and classical physics with its arbitrary energy values is obtained as a limit. Mathematically, this is a very economical way of showing that nothing of the old classical theory is lost. Statements of the latter refer only to the subset of situations in which the discreteness of energies is irrelevant anyway.

How classical behavior is approached can even be demonstrated explicitly for suitable wave functions. This is important because there are, despite quantum theory's universal applicability, many situations in which macroscopic objects are still described very precisely by classical calculations. A goalkeeper has to pay attention to many things during a penalty kick, but need not worry that the ball, as a consequence of the uncertainty principle, will suddenly appear at a position completely different from where it was just a moment ago. Wave functions must exist for macroscopic things that do not noticeably exhibit the strange quantum phenomena. Such wave functions are called *half-classical* or *semiclassical*, for although they are indeed quantum theoretical objects, their behavior is largely explicable by classical physics.

An atom's ground state is not of this kind, because it shows strong quantum behavior; otherwise it could not solve the classical problem of instability. Similarly, the near excited states with slightly higher energies, in which quantum jumps are still sizable, are not semiclassical. But when atoms go to higher energies, thus stronger excitations, quantum jumps become milder. This must happen because there are infinitely many energy levels, but only a finite amount of energy is required to remove the electron from the atom. Indeed, atoms can be ionized by adding a sufficient amount of energy, removing one or more electrons from the orbits to leave a positively charged ion behind. All the infinitely many energy levels occupy a finite energy range and must crowd together at some point (figure 6). This is the place where distances between neighboring energy levels become very small, a behavior much closer to the classical one.

Wave functions for an atom in this dense energy range can easily behave classically. As in figure 8, such a wave function can be imagined as a single crest with an extremely sharp top. In time, it follows a trajectory such as the classical equations of motion would have determined. What we have here is a wave function of a certain

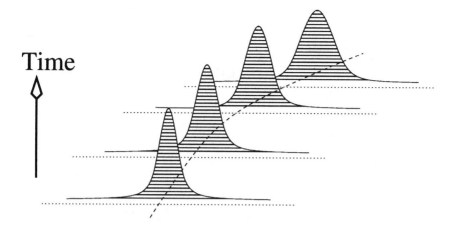

Time

8. A wave moving along the dashed curve, spreading out in the course of time (upward).

spread; strictly speaking, the electron's position is still not precisely defined, but if we do not look too closely, the extension of the wave function can, after a first approximation, be ignored, allowing us to view the wave's peak position as the "location" of an electron. In this way, the classical theory presents as a limiting case of the more generally valid quantum theory.

If we do look closely, even semiclassical waves do not exactly follow the classical laws of motion. Watching the motion of a semiclassical wave function over longer time intervals, we will notice deviations from the classical trajectory, first small but then becoming more and more pronounced. The wave function as a whole is, after all, subject to temporal evolution as given by the Schrödinger equation (again, a differential equation for the change of the wave function during the smallest changes of time). Not only the peak position of the wave but also its width and shape matter. Only in exceptional cases can the peak follow the classical trajectory exactly; otherwise it is yanked away from the classical curve by the changing shape of the whole wave function. As in the case of water waves, an uneven ground influences the velocity of various parts of the wave differently. On one side of the peak it may fall back behind the maximum, and it might seem as if the slope of the wave mountain will come sliding down. The mountaintop is made to move by the changing shape of the whole mountain, in addition to its rigid motion; both contributions are illustrated in figure 9. For a semiclassical wave

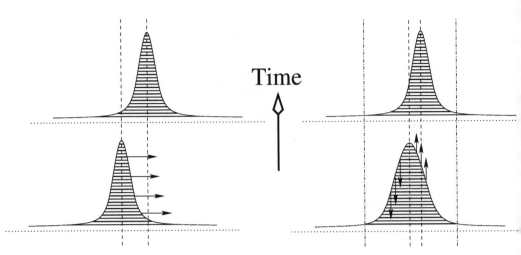

9. Depending on the ground, a wave can move rigidly without changing its shape (left), or it can change its height differently at different positions. The wave crumbles, for instance on its left-hand side, such that its maximum seems to move to the right. Such changes affecting parts of the wave do not occur in classical physics, and yet they have an influence on the position of the maximum. To describe the position precisely, quantum corrections to the classical equations in the form of effective forces become necessary.

function, such deviations caused by the shape are small at any given time. Initially, it may seem that the classical trajectory would be realized precisely. Before long, however, deviations do add up and corrections must be incorporated for an exact description. Thanks to those deviations from classically expected curves, effects of quantum theory can often be confirmed very precisely by experiments.

Provided that the corrections correspond to small shifts of the trajectory (in comparison, for instance, with the radius of an atom in such a state), one can apply mathematical approximations called *perturbation theories:* The deviation is seen as a small disturbance of the classical trajectory and can be computed with more ease. Instead of computing such perturbations directly from a semiclassical wave function, solving the rather complicated Schrödinger equation, the classical equations themselves can be modified. If the mathematical procedure is followed correctly, solutions to the new equations describe the motion of a wave function better than do solutions of the purely classical equations. Corrected equations can, for instance, incorporate a force unknown to the classical theory. Such a quantum force would constitute the effective description of implications of the changing wave mountain on its peak position.

Modifications to the classical equations are called *quantum corrections,* and in contrast to the classical terms, they depend on Planck's constant. When this constant is set to zero, quantum corrections vanish—another example of how the classical theory appears as a limit of the quantum theory. Quantum-corrected equations are often very efficient and not only allow simpler computations of certain quantities, but often provide an intuitive understanding of quantum phenomena. While wave functions are indirectly used in the determination of quantum corrections, their properties, such as difficulties in simultaneous measurements of position and velocity arising from the uncertainty relation, need not be taken explicitly into account. All this is, after all, already provided for in the derivation of correction terms. Such equations—called *effective*—have played a large role ever since the early stages of quantum mechanics.

Especially in semiclassical applications of quantum theory, one can often find effects for which, at least to some degree, visualization by analogy with water waves on a lake applies. A wave's maximum corresponds to the position of a classical particle, but the wave must always be spread out: The particle's position is not determined precisely. Because of the spreading, a wave can stretch out its sentinels and become sensitive to possible unevenness in different areas of the lakebed, even at places where the particle would never tread in classical physics. Tails of the wave can, for instance, extend into more shallow parts of the lake even if the maximum lies far away. A wave propagates differently in shallow areas than in deep ones: The speed of waves depends on the depths of the lake where they are traveling. Accordingly, the direction of the whole wave as well as its form change depending on how the lakebed is formed. A well-known example is the fact that ocean waves always move toward the beach because they are redirected by the depth decreasing toward the land. Changes in the shape of a wave due to the ground profile can be observed in an impressive way by the piling up and eventual breaking of waves near the beach.

Breaking waves show strong unparticlelike effects and can no longer be considered to result from mere perturbations of simple motion. But before it comes to this, the seabed's profile causes the peak position to shift only slightly, which can be taken into account by a correction to motion over level ground. Moreover, the wave will spread out ever more, sending its sentinels farther in new areas: Owing to their different propagation speeds at different depths, separate components of an extended wave experience slightly distinct shifts over the course of time, and the wave becomes more dispersed. (In exceptional cases a wave may be concentrated into a small region, but this would require a fine-tuning of all constituent wave contributions. Moreover, even then the wave would, after having been focused at one time, disperse again.)

In addition to the need for corrections to a particle trajectory when a wave symbolizes the imprecise position, the picture of a water wave illustrates the long reach of quantum effects: Distant regions, which classically would play no role whatsoever, influence the wave position. An exact description requires knowledge of the whole wave, not only of its maximum. Also, the spread widens in time, and it influences the mean position. Aside from position and spread, many other parameters are necessary for an exact description of the waveform. No finite number of parameters will suffice to describe the complete wave precisely, for this would require knowledge of the height of the wave at all its infinitely many positions. All these parameters are, as a result of their mutual influence on one another, subject to complicated interplay in time. In general, a correct mathematical description can be obtained only by means of the Schrödinger equation and its solutions. Luckily, we are usually interested not in the whole wave but only in a few parameters such as the peak position and the spread. For not too large a spread, for instance in semiclassical states, the set of infinitely many numbers can be reduced to a handful. Behind this reduction lies the mathematical procedure of effective equations.

Such approximation techniques pervade all of physics. They can be found in particle physics, in which complicated interactions of elementary particles can, at moderate energies, be described by effective equations. In high-energy physics, instead of the spread of a wave function, one commonly speaks of the creation of a particle-antiparticle pair—a process based on another consequence of the uncertainty relation: Energy, too, is imprecise and can, for brief periods of time, be used to create a pair consisting of a particle and its antiparticle—on the condition that the pair must annihilate itself shortly to return the borrowed energy. Here, the wave function is being spread out when suddenly two new particles appear.

Just as the propagation and spreading of waves on a lake are influenced by the lakebed, the process of pair creation depends on the ground—here, the form of space-time. Astonishing implications will be seen later in cosmology and in black hole physics. Sometimes pair creation can give rise to drastic phenomena, comparable to the buildup of tsunamis after seaquakes, when rapidly moving ground generates high waves on the surface. When space-time expands, especially in an accelerated manner, it can generate particle waves taking the form of new matter. In contrast to the seabed during a quake, the universe does not vibrate periodically, and consequences are thus not too dramatic. But what is important is the acceleration of the ground. As we will see in chapter 5 on observational cosmology, accelerated expansion, called *cosmic inflation,* is indeed assumed to have taken place in the very early universe. If this idea is correct, the accelerated expansion active at those times just sufficed to create all the matter now present, like a wave piling up out of a vacuum.

So far, we have looked only at a single wave in its entire extension. But in the universe, as on a lake, there are usually many waves superposed on one another. In such a superposition the waves influence one another's motion even without direct contact, in contrast to hard classical particles. If waves should grow far apart, they can still retain a memory of the interaction for a long time—a phenomenon called *entanglement,* following Erwin Schrödinger. In quantum mechanics, this even goes so far that an event happening to one wave—such as breaking on the shore, or a quantum mechanical measurement process—can strongly influence other waves far out in the ocean. For sea waves, such a strong effect would contradict expectations, but in quantum mechanics it is possible due to the so-called nonlinearity of the measurement process: Small changes in a nonlinear system can have disproportionately strong effects throughout the system.

But why, then, are macroscopic objects such as we find in everyday life not entangled in this way? In such cases, after all, direct contact, or at least the transmission of a signal, is usually required in order to have any influence. It turns out that the entanglement of wave functions reacts very sensitively to tiny perturbations that we usually do not even notice. Just single air molecules or light or (in a dark vacuum) heat influences the wave functions so strongly that they, as if in sensory overload, forget their previous entanglement in a process called *decoherence.*

Here, too, the image of a lake may be helpful, but now of a lake in the rain. Large waves are still visible and can be tracked in their propagation for long times; but every single raindrop contributes a new little wave. In this sea of irregular perturbations, details of the entangled interrelations drown. Similarly, the world on a large scale looks classical because fluctuations and entanglement are easily destroyed. On the small scale, however—in microscopic physics or in precisely constructed experiments—perturbations can often be subdued long enough to let the unfamiliar world of quantum physics be revealed.

Thus Schrödinger's cat is at last delivered from uncertainty: In any life-form, so many processes take place that quantum mechanical superpositions are always perturbed. Even if no measurement is performed on the apparatus or the cat, the superposition of decay and nondecay of the radioactive substance—the life and death of the cat—is rapidly transformed to a definite state. For some time, the decay probability is very small, owing to the weakness of the radioactivity, and the cat remains alive. Strictly speaking, the atoms in the substance are indeed in a superposition, but the contribution from a state of decay is very small. Perturbations of the system quickly and most likely turn the superposition into a definite nondecay state. But at some time, the probability of decay becomes large enough to make the superposition register as a definite decay, and the cat must suffer its fate.

BLACK BODY RADIATION:
THE SIMPLICITY OF WEARING BLACK

In addition to the emission and absorption spectra of atoms, the measurement of heat radiation in a closed dark box at a uniform temperature, the so-called *black body*, was an important subject of experiments and theoretical modeling in the early years of quantum theory. What we perceive as heat are tiny vibrations of atoms and molecules. These constituents, being electrically charged, radiate energy when vibrating: heat radiation. Radiation in a closed box can be controlled very well, and precise data were collected in the course of the nineteenth century.

Initially, these observations were consistent with classical ideas about radiation. In particular, experimenters had measured how the radiation energy is distributed over different frequencies. With increasing frequency, energy should grow, since atoms and molecules emitting radiation vibrate more strongly. If one has seen a certain amount of energy realized for one frequency, there must be more energy in every higher frequency. Indeed, this increase was measured—but it also led to a serious problem in understanding: According to classical calculations, the energy should grow without limit at increasingly high frequencies; at sufficiently large frequencies, the energy would exceed any conceivable upper bound. In such a way, the total energy, summed over all frequencies, becomes infinite. Here we encounter a problem in some ways resembling the infinite increase of energy at cosmological singularities or in the interiors of black holes. Black body radiation and its quantum theory are much better understood and much more scrutinized experimentally than quantum gravity; it is thus instructive to see how the problem of infinite energies is resolved here.

As discovered by Max Planck even before quantum mechanics was developed, the solution to the problem is the discrete nature of energy emission. The atoms and molecules of the black body walls have discrete emission spectra and cannot emit arbitrary energy packets. Taking this into account—something Planck did intuitively even without knowledge of the underlying quantum physics—the infinite increase of energy at high frequencies is prevented, and a finite total energy results. Historically, Planck relied mainly on precise data already showing slight deviations from the expected classical behavior. Planck's formula, on the other hand, agreed exactly with these new observations. In 1905, it was Einstein who explained, in another groundbreaking publication, Planck's formula by the existence of a discrete distribution of energy in heat radiation. He was thus the first to introduce quantized particles of light, or *photons.*

Planck's formula shows that the energy distribution of matter at high frequencies—corresponding to small wavelengths of radiation—behaves differently from what is expected with classical understanding. Like the stabilization of atoms, this may have consequences for a complete description of quantum gravity, for it is energy that, according to relativity, is equivalent to mass and thus gravitationally active. If energy behaves differently at small wavelengths or small distances than expected in the (non-quantum) general theory of relativity, a modified form of space-time as it is determined by matter should result. Space-time curvature and the gravitational force it implies would differ from its form in classical solutions. Such extreme scales would be significant especially at the big bang, when the universe is very small, or in the collapsed centers of black holes. If the quantum theoretical gravitational force were no longer purely attractive in these regimes, an infinite increase of energy could perhaps be prevented, and stable situations thanks to repulsive gravitational forces would result. Whether this is the case can be answered only with a concrete quantum theory of gravity. But it is encouraging to see that two independent consequences of quantum theory—the stability of atoms and the finiteness of black body radiation—both suggest this possibility.

PLANCK SCALES:
ON THE ENORMITY OF BEING QUANTUM

Before turning to possible quantum theories of gravity after the next chapter, we should take a more direct look at the scales to be expected. So far, as already mentioned, general relativity agrees very well with observations. Nobody would expect quantum gravity to be needed anytime soon to explain directly observed phenomena. In contrast to the numerous practical problems before quantum mechanics was developed, the need to develop a theory of quantum gravity is of an entirely conceptual nature, motivated by, for instance, the desire to avoid the singularity problem. While similar types of problems did occur in the classical physics of matter before the development of quantum mechanics, they were marginalized by the wealth of concrete observations.

The role played by quantum gravity for observations can be understood by looking at the relevant length scales, in particular the

Planck length (not to be confused with Planck's constant). Although Planck was no quantum gravity researcher, he had already introduced this important quantity.[3] He had noticed that the speed of light, Newton's constant for the gravitational force, and the later named Planck constant taken together allow, by suitable multiplications and taking roots, the definition of a length: the Planck length. By different combinations one can also obtain a time, the *Planck time,* and a mass, the *Planck mass.* This looked fascinating because no actual length measurement is required to determine the speed of light, the gravitational constant, and Planck's constant—and yet one obtains a unique length by simple mathematical combinations. Possibly, this connection could be exploited for sensitive measurement methods or a new reliable convention for the unit of length.

So far, these thoughts have not played a technological role, due in part to the extreme smallness of the Planck length. Compared to the radius of an atom of about 0.00000001 meter (a billionth of a meter) or even that of a nucleus of 0.000000000000001 meter (a quadrillionth of a meter), the Planck length is infinitesimal. Inserting the known values of the speed of light and the gravitational and Planck constants, one obtains a value of about 0.000000000000000-0000000000000000000001 meter (ten millionth of a quadrillionth of a quadrillionth of a meter). As many orders of magnitude lie in the ratio of a meter to the smallest currently measurable distances—the size of elementary particles seen in accelerator experiments—as there are in the ratio of those smallest measurable lengths to the Planck length. If we could magnify the universe so as to make a region of Planck size as large as an electron, an electron would be enlarged to the size of a football.

To obtain the value of the Planck time, we can think of it as the time required for light to pass the distance of one Planck length. This is clearly tiny; the number of Planck time intervals in one second has 42 digits. The Planck mass may sound more down-to-earth, at about a microgram. But this value is also extreme if we think of it as a fundamental mass, such as the mass of an elementary particle rather than of a composite object made from a gazillion constituents.

How extreme the Planck scales are is perhaps best indicated by the size of the *Planck density:* the density corresponding to a Planck mass in a cubic region of edge length given by the Planck length. This density is the equivalent of a trillion suns compressed to the size of a proton. Since the Planck length contains mainly Planck's constant, which is crucial for quantum theory, and the gravitationally important Newton's constant, one expects its magnitude to play a large role in a combination of those two theories. In cosmology, quantum gravity becomes indispensable when the universe near the big bang singularity is compressed to Planckian density. Such dimensional arguments are often reliable, at least approximately, and they provide important expectations for theories still to be developed. For instance, the radius of a hydrogen atom in quantum mechanics can be estimated in a similar way because the electron's charge and mass, the defining properties of the system, together with Planck's constant, which is required by the quantum nature of the problem, determine a length parameter in a unique way. Up to a factor of two, this length agrees with the quantum mechanical extension of an electron's wave function in the ground state of a hydrogen atom, the so-called *Bohr radius.*

We can easily see why no observations exist so far that would without a doubt require a quantum theory of gravity. After all, the Planck length is much, much smaller than any length measured thus far. Should it one day be possible to resolve lengths about the size of the Planck length, quantum gravity and the space-time structure it implies would become important. Then again, such arguments are to be taken with a grain of salt, for they fail in more complicated situations in which several distinct parameters are involved. Sizes of heavy atoms such as uranium, with a large number of electrons—in this case, 92—cannot be estimated by simple dimensional calculations. Would we have to multiply the Bohr radius by the number of electrons to obtain the radius of a uranium atom? Or divide by it? Or perform some other procedure? More detailed knowledge of the physics at play is needed, a full theory.

Moreover, different effects can, as we will describe later, result in magnifications by adding up many tiny terms. Even though quantum gravitational deviations from general relativity should, at a fixed time during cosmic evolution, be tiny, such corrections could be enhanced in the long duration while the universe expands; they would become detectable sooner than the smallness of the Planck length might indicate. Such indirect effects have often played large roles in the discovery of new physical phenomena. A well-known example is Brownian motion, used to uncover the atomic structure of matter, in which the tiniest impacts of molecules in a liquid cause a vibrating motion of suspended pollen grains, visible by light microscopy. We will return to this phenomenon in the chapter on cosmogony (chapter 8).

Most important, the value of the Planck length, which is very small but, after all, nonzero, shows that some cosmological situations such as the big bang can only be understood with a quantum theory of gravity. If the whole extension of the universe is supposed to vanish at that time, even the Planck length would be gigantic by comparison. With such a theory one could analyze the singularity problem and the possible existence of stabilizing repulsive forces. While no completely formulated theory yet exists, there are several concrete expectations and many encouraging signs.

3. AN INTERLUDE ON THE ROLE OF MATHEMATICS
THE WORLD IN AN EQUATION?

I had to learn thus with a frown:
No thing can be where words break down.

—STEFAN GEORGE, *The Word*

General relativity and quantum mechanics, taken separately, result in numerous confusing phenomena or even apparent paradoxa, not just for inexperienced observers but also for longtime researchers. A clear view is possible only thanks to their mathematical formulation, without which evaluating and understanding these theories would be impossible. It takes significant effort to learn and develop these theories—in particular when one aims at quantum gravity, in which the underlying mathematical concepts of general relativity and quantum mechanics are to be combined. Once learned, however, mathematics is incorruptible and lays out strict recipes to answer questions on which Intuition would refuse to take a stand or would even entangle herself in paradoxa.

Ever since Galileo Galilei, the mathematification of physics has been responsible for its unprecedented success. It leads us to entirely new worlds—allegorical worlds of new phenomena, literal new worlds in cosmology, or "brave new worlds" with a critical or prescriptive touch. According to Galileo, mathematics is the language of nature, indispensable for its understanding. Mathematical formulas are nature's sentences, and its words are mathematical variables and operations. "No thing can be where words break down": Without mathematics, scientific knowledge of nature is confined to objects directly accessible by the senses. At the singularities of space-time as they occur in general relativity, things, even entire worlds, are literally lost. The breakdown of the most important sentence of the theory, Einstein's equation, prevents any access to the world before the big bang or deep into the heart of black holes. New mathematical vocabulary is still to be coined.

Mathematically predicted phenomena bring a new quality into physics. On the one hand, physical theories can explain sensory perceptions, such as the aurora borealis or the blue of the sky. But theories also give rise to new impressions experienced only abstractly, thus extending the realm of our senses. Examples can be found especially in cosmology, in the phenomena of the big bang or black holes. Such impressions exist only in the mathematics underlying the theories, but they must be tested by observations in the real universe even if those tests may be indirect rather than based on direct sensory imprints. It is this combination of strict mathematics and undeluded confirmation in nature that elevates physics above mere speculation or myths. Only this makes it trustworthy.

MATHEMATICS FOR ITS OWN SAKE?
MATHEMAGIC

> Pfuehl was one of those theoreticians who so love their theory that they lose sight of the theory's object—its practical application. His love of theory made him hate everything practical, and he would not listen to it. He was even pleased by failure, for failures resulting from deviations in practice from the theory only proved to him the accuracy of the theory.
>
> **—LEO TOLSTOY,** *War and Peace*

In recent times, mathematics has found a life of its own in physics—especially in quantum gravity research, which is not yet subject to policing observations and in which mathematical consistency serves as the only selection criterion by which to choose theories. This is indeed a strong condition in quantum gravity, whose high level of difficulty does not allow many consistent formulations; so far, after all, no completely satisfactory form has been found. On the other hand, consistency conditions are usually difficult to evaluate, and so the concept of beauty as an additional criterion often enters the stage. Beauty is easier to spot than consistency, for it is based on an emotional process even for theoretical physicists, who are often seen as rather prosaic. However, mathematical beauty does not reveal itself to the untrained eye. More critically, even in mathematics, beauty is in the eye of the beholder; and so various research directions emerge, all following different ideals and often communicating with each other only with difficulty. Above all, one loses sight of reality: "There is no doubt, the veracious one, in that bold and final sense as the belief in science requires it, affirms *therewith a world different* than that of life, of nature, and of history; and insofar as he affirms this 'other world,' what? is he not, by just this, led to its counterpart's, this world's, *our* world's— negation?"[1]

Once one arrives at such an intimate relationship with mathematics, one can easily confuse mathematical objects with reality. John Stachel, in this context, speaks of the "fetishism" of mathematics: "I designate as the 'fetishism of mathematics' the tendency to endow the mathematical constructs of the human brain with an independent life and power of their own."[2] Occasionally, the words already indicate what criteria (scientific or otherwise) are used to develop certain theories. In some branches of quantum gravity research, the language often appears overblown, manifesting itself, for instance, in descriptions of mathematical facts as "magical" or "mysterious."[3]

In all these considerations, it is important to remember what the aim of science is: to describe Nature; and Nature has her own sense of beauty. Nature alone can decide which theories best describe and predict her behavior. What too strong an insistence on human concepts of beauty can lead to in the exploration of Nature was Faust's experience in the following passage and its continuation:

SPIRIT: Who calls me?
FAUST: *(turning aside)* Dreadful shape!

SPIRIT: With might,
Thou hast compelled me to appear,
Long hast been sucking at my sphere,
And now—
FAUST: Woe's me! I cannot bear thy sight!
SPIRIT: To see me thou dost breathe thine invocation,
My voice to hear, to gaze upon my brow;
Me doth thy strong entreaty bow—
Lo! I am here! —What cowering agitation
Grasps thee, the demigod! Where's now the soul's deep cry?
Where is the breast, which in its depths a world conceiv'd
And bore and cherished? which, with ecstasy,
To rank itself with us, the spirits, heaved?
Where art thou, Faust? whose voice I heard resound,
Who toward me press'd with energy profound?
Art thou he? Thou, —who by my breath art blighted,
Who, in his spirit's depths affrighted,
Trembles, a crush'd and writhing worm!
FAUST: Shall I yield, thing of flame, to thee?
Faust, and thine equal, I am he!
SPIRIT: A constant weaving
With change still rife,
A restless heaving,
A glowing life—
Thus time's whirring loom unceasing I ply,
And weave the life-garment of deity.
FAUST: Thou, restless spirit, dost from end to end
O'ersweep the world; how near I feel to thee!
SPIRIT: Thou'rt like the spirit, thou dost comprehend,
Not me!
(*Vanishes.*)

—**GOETHE,** *Faust*

Turning away from Nature, symbolized by the Earthly Spirit, did have unfortunate consequences not only for Faust, who afterward made a fateful pact with Mephisto and his magic, but also for several others initially uninvolved. Faust gains more and more power, but loses sight of the essential.

ABUSE OF INFINITY:
THE DIRTY TRICKS OF PHYSICS

The moving moves neither in the space where it is, nor where it is not.

—ZENO OF ELEA, Fragment

The concept of infinity is a dangerous weapon in certain mathematical arguments, and so it comes as no surprise that it is sometimes abused. As the discoverer of the power of infinity we may identify Zeno, who, as a student of Parmenides of Elea, tried to support his teacher's ontological theory. (In addition to this theory, Parmenides had provided explanations for the phases of the moon as well as of other astronomical phenomena, which makes him one of the first important cosmologists.) Zeno's problem was the boldness of Parmenides' theory, which claimed that no motion exists and that any perception of motion is illusion. In the defense of this theory, Zeno employed several arguments, in their modern formulation all based on the concept of infinity. He constructed, for instance, a footrace between Achilles and a tortoise. As a fair sportsman, Achilles grants his weaker competitor a head start. Once he begins to run, Achilles quickly reaches the tortoise's starting point, which in the meantime has inched forward a bit. Achilles needs some time to get to this place, but by then the tortoise has again crawled ahead. Repeating this process over and over again, Achilles never catches up with the tortoise. If Achilles cannot overtake the slow-moving tortoise, in contrast to our expectation for motion, motion itself can only be illusion.

Zeno tries to ensnare us. He certainly knows that Achilles will catch the tortoise after a certain amount of time, for he surely had watched runners compete, or even competed himself. (He was, perhaps, being petty enough to attempt a mental revenge for a defeat.) Now he subdivides the time between start and catch-up into infinitely many smaller intervals, changing single-handedly, but only mentally, the flow of time. Instead of allowing time to run as usual, he jumps from one interval to the next. With intervals becoming shorter and shorter, time in his thoughts passes differently; it is slowed down more and more compared to the conventional flow. He thus transforms a finite range—the time required for Achilles to catch up with the tortoise—into an infinite one. Mathematically speaking, he performs a coordinate transformation, mapping the finite time of the catch-up to an infinite value. His argument takes place in the new time, in which the infinite value is indeed never reached. But he overlooks, or tries to seduce us into doing so, the fact that in our actual perception of the competition only the original, conventional time is significant, which requires just a finite period for the tortoise to be overtaken.

The concept of infinity is dangerous, but often too enticing to resist. Especially in moments of Zenoic desperation—situations in which one despondently tries to construct a proof in the face of better knowledge—infinity is time and again abused in physics. Comparable thoughts can, for instance, be found in the context of our main problem, that of singularities in general relativity. Many of the proposed arguments can be reduced to a transformation of time, mapping the finite duration between the big bang singularity and the present to an infinite one. Viewed in this time, the big bang happened an infinite amount of time ago, thus at no finite point in time, or never. What is overlooked here is, of course, that the newly introduced concept of time, which plays only a mathematical role, is irrelevant, unlike physical (also called proper) time as perceived by us. An astronaut falling into a black hole would hardly be consoled by the fact that his small, finite remaining lifetime can mathematically be mapped to an infinite range.

ON HYPOTHESES AND THEORIES:
CODE OF CONDUCT FOR A THEORIST

I have not been able to discover the cause of those
properties of gravity from phenomena, and I frame no
hypothesis. . . . It is enough that gravity does really exist,
and acts according to the laws which we have explained.

—ISAAC NEWTON

Despite all seductions, mathematics cannot be banned from physics. Depending on the degree of mathematical formulations, one distinguishes between *hypotheses* and *theories*. Hypotheses are of speculative character and usually mark the beginning of investigations in a new range of problems and the buildup of ideas. Theories are, in contrast to the vernacular use of the term, of much higher maturity; they are supported by many independent tests as well as, ideally, by rich observations. Theories are the foundation of physics; they bring a plethora of observations into systematic order, explain phenomena such as the expanding universe through physical images, and even risk new predictions. The most important statements of theories are knighted by bestowing on them the title of "natural law." This does not mean that those laws and the theories containing them will be valid forever, for one can never rule out new observations that cannot be explained by any yet existing theory. But successful theories describe the overwhelming majority of observations already made; they are the key to the order of known physics.

It is initially mathematics that raises the concept of theory in physics to a much higher status than is granted colloquially. Theories are based on simple principles readily accepted by experienced physicists, but they imply much less obvious statements, often following only after long calculations.[4] Such implications, in contrast to underlying principles, can be subjected to experimental tests. If these are passed successfully, the theory is established; otherwise it will soon be forgotten. In this sense, special and general relativity are indeed theories.

Hypotheses, by contrast, are much less secured and often remain to be completed by a systematic theory—unless they are disproved in the process. A rather influential hypothesis, held throughout the nineteenth century but eventually disproved, was the postulate of the ether, a pervading medium thought to support the wavelike excitations that we perceive as light. As a consequence of this hypothesis, light rays moving in various directions on Earth—all aimed differently with respect to Earth's motion through the space-filling ether—should move at different speeds. Several experiments were undertaken to detect such velocity changes, and thereby possibly provide insights into the nature of the ether. Finally, Albert Michelson and Edward Morley were able to conclude in 1887 that light moves with the same speed in all directions, disproving the concept of the ether. A short while later, this observation found a natural explanation within special relativity. A successful example is the light particle hypothesis, referred to thus even by its originator, Einstein. In those days, around 1905, the atomic composition of matter was not accepted at all; assuming the existence of photons as particles of light must have appeared objectionable to many. Even though Einstein was able to derive Planck's formula for black body radiation from this hypothesis—a considerable success—the assumption itself remained to be secured further. Only much later did this hypothesis mature into a theory, culminating in modern *quantum electrodynamics* as the theory of photons and their interactions with electrically charged matter.

What is nowadays referred to as quantum gravity is, strictly speaking, not yet a theory even though many independent (but only mathematical) tests have successfully been done. The final demonstration of complete consistency is missing, and so far not a single supporting observation is available (though no observations clearly contradicting such theories exist either). Even so, these edified thoughts are more than just hypotheses, and for simplicity's sake they are often called theories—or sometimes, a little less boldly, called *frameworks:* The frame is clear, but much of the interior remains to be filled in. All the work done so far in quantum gravity, including what is discussed in this book, is, as a scientific formulation, situated between a hypothesis and a full-fledged theory; the field must still be considered speculative. Only observations, possibly of a kind described later, will be able to change this status.

4. QUANTUM GRAVITY
COMBINING EVERYTHING

When Paul Dirac was working on a combination of special relativity and quantum mechanics in the years leading up to 1928, he was driven neither by serious conceptual problems nor by an erroneous theoretical description of experimental data. He had realized that these two theories—Einstein's special relativity and Schrödinger's formulation of quantum mechanics—were very successful in their own respective domains, but unfortunately incompatible with each other. Schrödinger's equation, which describes the temporal evolution of a wave function, does not take into account the relativistic transformability of space and time, recognized by Einstein and by then reliably established through observations. A full, consistent view of physics cannot tolerate such a coexistence of incompatible theories, and so Dirac set out to reformulate the Schrödinger equation relativistically.

In physics before special relativity, the kinetic energy of a moving body is proportional to the square of its velocity. Every driver knows this, for the distance needed to stop a car by transforming its kinetic energy into heat in the brakes increases with the square of the velocity rather than linearly. Thus the consequences of a crash at a speed of 70 mph are, unfortunately, much more than just twice as devastating as those of a crash at a speed of 35 mph.

In special relativity, by contrast, energy behaves like this only at sufficiently small velocities. Otherwise, one has to take into account that mass as well as motion contributes to the energy, according to Einstein's famous formula equating energy with mass multiplied by the speed of light squared. Moreover, energy and velocity (or, more precisely, momentum as the product of mass and velocity) play mutually transformable roles similar to what we have already seen for space and time. Energy as well as velocity thus enters the bottom line of energetic processes, such as stopping a car, with their squares.

Schrödinger's equation makes use of the prerelativistic relationship and disregards the square of energy. Dirac's new equation, on the other hand, does take into account the relativistic relation. But the square of a number, in contrast to the number itself, is independent of the sign: Minus one times minus one is one, just as is one times one. For every solution of the Schrödinger equation there are always two solutions to Dirac's equation, differing in their energy signs and possibly in that of charges, too, but otherwise agreeing in properties such as the mass. One of the unshakable (and well validated) beliefs of theoretical physics is that every solution to a theory must either be inconsistent with some mathematical principle or correspond to something real. Dirac thus predicted the existence of a new world of phenomena of matter, just by mathematically combining special relativity and quantum mechanics: For every known particle, such as the electron, there should be an antiparticle of the same mass but opposite charge. When Dirac published his equation, such particles had not been observed; Dirac's prediction thus came at high risk. (He reportedly hesitated to state it in clear terms.) But soon afterward, in 1933, Carl Anderson provided a direct detection of antimatter in the form of positrons (the antiparticles of electrons), a feat repeated later on for the other known particles. By now, corrections implied by the Dirac equation compared to Schrödinger's can be measured very precisely, for instance in the frequencies of light emitted or absorbed by hydrogen.

Dirac arrived at this far-reaching view of a new world of matter by nothing more than a mathematical analysis of combining two theories. Nowadays we are faced with a similar, though mathematically much more challenging, problem of combining general relativity and quantum theory. What new worlds will such a combination reveal? Special relativity determines the motion of particles in space-time, while general relativity describes the behavior of space-time itself. It is then natural to expect a consistent combination of general relativity and quantum physics to reveal not new matter but new regions of space-time, or new parts of the universe.

To understand space-time and the universe completely—including regions which general relativity can only represent as singularities—we desperately need a quantum theory of gravity. Only then can we obtain a reliable description of high-density states such as the big bang or black holes. A complete theory, alas, is not available; rather, there are several very different approaches, all having distinct strengths and weaknesses. Even in their basic principles they strongly deviate from one another. The reason for this diversity is that the mathematical foundations of general relativity and quantum theory, as well as the understanding of their concepts, can be combined in many ways. At present, we don't know which of the proposed combinations is the correct one—or whether we will have to look for yet another route to reach the goal.

APPROACHES TO QUANTUM GRAVITY:
STRENGTH IN DIVERSITY

For a wide, experimentally untested field such as quantum gravity it is not surprising that vastly different routes to a potential completion are being taken. A count of researchers and publications indicates that the most-traveled roads are *string theory* and *loop quantum gravity,* which are indeed, as we will see, very different from each other. In addition to these two, intriguing quantum-gravity-related proposals such as twistor theory, noncommutative geometry, causal sets, and causal dynamical triangulations are being developed by smaller numbers of researchers, where progress is correspondingly slower. Many of the key ideas in quantum gravity can already be illustrated by string theory and loop quantum gravity, and so we will focus on those.

STRING THEORY: THE RICHEST SYMPHONY

String theory has attracted the greatest interest. Having started from older developments in particle physics, its strengths rest in particular on a quantum theoretical description of excitations such as gravitational waves on a fixed space-time stage. In string theory, there are particle-like objects, *gravitons,* which transport the gravitational force in packages just as photons carry light. But this is almost a side effect, for the basic concept of string theory is in fact much more radical, as it proposes to leave behind the fundamental picture of particles as pointlike or as hard solid balls with tiny extensions. Even the indistinct but still somewhat localized wave function of quantum mechanics is surpassed by the string concept.

Instead, particles such as gravitons as well as the constituents of matter—the electron, the quarks forming protons and neutrons, and additional elementary particles generated at high energies of accelerator experiments—are understood as excitations of one single elementary object: the string. As the string of a musical instrument can create diverse sounds by differently excited vibrations, a fundamental string can oscillate in many ways. Just as sounds are distinguished from one another by different frequencies, the vibrations of strings have varying energies or masses. In principle, they could explain the observed masses of elementary particles, if the calculated numbers agree with the precisely measured values known from accelerator experiments.

By reducing all phenomena of particle physics to a single object, string theory promises to unify all the known fundamental forces— gravity, electromagnetism, and the strong and weak interactions—in a single force formula. There would no longer be different concepts such as space-time as the carrier of gravity and the electromagnetic field as the messenger of the electric force, but just a single object from whose vibrations would emerge all forces and the matter particles on which they act. This object in its elementary form is the eponymous string.

Unifications of different theories and forces have played major roles in the development of physics, for instance when Maxwell combined the initially unconnected electrical and magnetic phenomena in one theory of electromagnetism. From such combined theories spring, almost inescapably, predictions of new phenomena that might be exploited technologically, or that can be used for independent tests of the theory. An example of applications of Maxwell's unification of electric and magnetic forces is the possibility of electromagnetic waves. These waves rely on an interplay of electric and magnetic excitations, requiring a unified theory of both forces: An electric field changing in time generates a magnetic field, and a magnetic field changing in time generates an electric one. Thus a changing excitation, for instance the current in an antenna, can enter into space when the emerging electromagnetic wave shifts from electric to magnetic energy, back to electric, and so forth. Without a combination of electric and magnetic excitations this would be impossible, and we would be denied applications such as radio or X-ray technologies.

String theory now attempts to unify all forces and particles by reducing them to a single string-object. There are certainly many different ways to construct a musical instrument based on a vibrating string. Differences of tone color can be discerned even by inexpert ears. Similarly, different string theories, all having different excitations, can be constructed in physics. Masses or interactions of particles as predicted by differently constructed theories would be distinguishable with detectors of high-energy physics, such as the Large Hadron Collider (LHC) at CERN near Geneva, which was launched in 2008 (and relaunched in 2009). Often, differences can be so drastic that one would not even have to do an experiment to rule out the theory: The stability of atoms depends on the masses of protons and neutrons, sharply confining the set of theoretically possible values.

Music arises from the varying tones of different instruments. Theoretical physicists, more prosaically, are interested primarily in simplicity and efficiency in their descriptions of nature. A plethora of theories would be unwelcome; unity, as demonstrated by Maxwell's exemplary electromagnetism, is preferred. String theory follows this ideal in an impressive way, not only in the potential power of its laws but also in their extremely elegant mathematical derivations. Here we do not have a whole orchestra of differently stretched strings, but only the soloist string theory itself. As it turned out after long years of research, all possible setups of fundamental strings are mathematically related. Various tone colors do not result in different physics; they are just different mathematical views of the same physics.

This statement is of unprecedented generality: All physical phenomena in gravity and particle physics could be described by a single theory without granting theorists any more freedom. Indeed, in the context of string theory one often uses the term "theory of everything" (TOE). Given sufficient control over the required mathematics, everything of interest could be computed in order to employ it subsequently for experimental tests of the theory. In principle, the theory can then be used to predict new phenomena, for instance at the time of the big bang.

In contrast to Maxwell's theory, experimental tests (or just test possibilities) as well as technological applications are still to be developed for string theory. One of the reasons is the complexity of its underlying mathematics. On the one hand, this complexity is the basis for the theory's uniqueness and its strong attraction for mathematicians and physicists alike; on the other hand, string theory appears as too grand a construct without (at least so far) being of much practical use. There are instead predictions of a rather worrisome nature concerning this theory's usefulness. For starters, mathematical consistency requires that the theory use more than three spatial dimensions; most often it requires nine. Only three of them—height, depth, and width—are certainly visible, explainable by tiny extensions of the remaining six dimensions. As a water hose may appear as a one-dimensional line when seen from afar, a nine-dimensional space with six tiny dimensions would look like a three-dimensional one. Only at close range would one notice the extra dimensions. Since quantum gravity effects are expected at distances near the small Planck length, additional dimensions suggested by a quantum theory of gravity could indeed be small enough to be unobservable. But other theoretical possibilities for an enlargement of such dimensions are being discussed that could have consequences for high-energy observations. Thus far, then, string theory, with its unification of forces, does indeed lead to new phenomena that could make the theory testable.

But the extra dimensions imply a serious problem. Even though the mathematical equations of string theory for high-dimensional space-times are complicated, one can estimate how rich the set of all its solutions might be. One can count those solutions whose properties agree roughly with what we see in our universe. Unfortunately, the result is an unimaginably large number: There are more solutions than there are protons in the whole universe. Such a large variety renders the theory, as beautiful and unique as it may appear from a mathematical viewpoint, useless for physical explanations. There is simply no basis for concrete predictions, since unknown phenomena would in no way be restricted. Such a theory would literally be a theory of everything, for everything—anything—could happen in it.

Even if a theory is unique and does not give rise to different tone colors as the vibrating strings of musical instruments do, this advantage can easily be obliterated by the vast deluge of solutions. There may be just one solo instrument, but its keyboard is enormous. Numerous pieces can be played on it, and nobody knows which composition corresponds to our world or just how this could be determined. Maybe there appear to be so many solutions only because all the conditions mathematically required for string theory have not yet been established with the available methods. Researchers split into two camps at this point: those who unquestioningly believe in the uniqueness of string theory and its solutions, and who continue looking for the missing consistency conditions; and others who, following the example of Leonard Susskind, make a virtue of necessity. If no unique solution is to be identified with our observed universe, they argue, at least the "most likely" solution should agree with our world. Of course, this brings in a new difficulty: first defining the degree of likelihood mathematically, and then, if this can be done reliably, showing how one would recognize the likelihood from the observation of only one of all possible worlds. We devote the last chapters to the uniqueness question of theories and solutions. For now, we turn to an alternative theory addressing the quantum theoretical nature of the space-time stage head-on.

LOOP QUANTUM GRAVITY: ATOMS OF SPACE AND DARKNESS

And weave the life-garment of deity.

—GOETHE, *Faust*

Matter is made up of atoms. If no atoms are present in an empty region of space-time, a vacuum state is realized. In modern physics, the vacuum plays an important role as the base on which all matter configurations rest. Starting with a vacuum region and moving matter into it atom by atom, all possible complex constructs can be built up. (This can be understood as a thought process, but with modern nanotechnology it is actually possible in a literal sense by manipulating single atoms.)

A material atom has energy that it contributes to the piece of matter it helps to build up. The energy can increase in two different ways: We can enlarge the number of atoms by adding more of them, and with them their energy; or an existing atom can be excited to a more energetic state. Practically, the second way of increasing the energy is normally done by heating the material or shining light on it. Since the total energy is conserved, atoms must be moved from one region to another but cannot be created from nothing, and the light or heat must come from an external source. All this is a direct consequence of the quantum theory of matter, whose language is *quantum field theory.* While quantum mechanics describes individual particles and their possible superpositions, quantum field theory includes the interactions of many particles.

In quantum gravity, we must apply quantum theory directly to space and time and their complicated games of change as they are realized in general relativity. According to loop quantum gravity, space is then also atomic. Spatial atoms are abstract rather than visual images. Some of their properties can best be described by attributing to them an extended, one-dimensional looplike rather than spherical shape, thus the name *loop quantum gravity.* In this atomic picture, the volume of any spatial region can only grow discretely in specific steps, by adding or exciting spatial atoms. As with energy, two different versions of growth are realized. But the process follows a different dynamic compared to material atoms since volume, unlike energy, is not conserved. The total volume can grow without having to be provided from an external source, with the ultimate consequence that the whole universe can grow in and of itself. In general terms, the interplay of space, time, and possibly matter in an expanding universe causes spatial atoms to be excited and occasionally, once they have become sufficiently large, to subdivide. Then a new spatial atom is formed.

A solid theoretical description needs a base to start from. One may view this quantum gravitational base as analogous to the matter vacuum, but it is much more ominous. In the absence of spatial atoms, no space exists at all, but it is still a physical state rather than nothingness. What is this state? It is distinct from empty space or the material vacuum, as is realized to a high degree in outer space. For an empty space, though devoid of matter, still has space: an extension and volume. There is not much to be found in the universe between the galaxies, but there is still something—an immense space. In loop quantum gravity, all volume, be it small or large, must come from spatial atoms; when we remove all of them, no volume remains. In an empty state of quantum gravity, not even space and volume exist. The vacuum of loop quantum gravity is of an inconceivable emptiness: no sound, no light, no stuff, no space; only time as a faintly glimmering hope to leave behind this wasteland.

One may view the spatial vacuum of loop quantum gravity in analogy to certain states in quantum theories of matter, but it does not correspond to the matter vacuum. Rather, even though such a state is devoid of anything, its more precise analog would be a state of infinite temperature. This observation led the quantum field theorist Klaus Fredenhagen to dub this state the *State of Hell.*

Any theory of quantum gravity fully addressing space and time must grapple with this State of Hell. As we will see later, this state is particularly relevant for the singularity problem, where space has collapsed completely. It is then not surprising that loop quantum gravity, as the most comprehensive theory yet of this utterly empty state, has made significant progress toward an understanding of singularities. It also shows that loop quantum gravity is much darker, and perhaps deeper, than the harmonious, playful string theory. Even large regions with many atoms of space keep a memory of the absolute vacuum, the State of Hell, they came from. There is a considerable theoretical risk: If the base state turns out to be incorrect, this original sin may bring down the theory.

COMPARISON: APPLES AND ORANGES

String theory and loop quantum gravity are unlikely allies that share an aim but have different ambitions and beliefs. Although they are scientific theories, beliefs enter in the form of principles. String theory is based on cherished concepts that have been derived through many decades of particle physics research, most important that of abstract symmetries as hidden relationships among different types of particles and forces. For instance, the electroweak interaction is based on a symmetry connecting electrons with neutrinos, two particles that at first glance are entirely different. Similarly, string theory proposes symmetries relating the gravitational force to the other known ones, such as the electromagnetic force. Loop quantum gravity, on the other hand, takes its principles from the long pedigree of the geometry of space, whose structure is considered more essential for an understanding of the universe as a whole than properties of the matter it contains. The first versions of such principles date back at least to the ancient Greeks, brought forward in modern form by mathematicians such as Carl Friedrich Gauss and Bernhard Riemann, then used crucially in Einstein's formulation of general relativity. For a quantum theory of gravity, it is a natural, though challenging, task to extend these concepts to an atomic space by combining them with elements of quantum theory.

Ambition is closely connected with belief, and it is not always easy to say which comes first. The dominant ambition of string theory is unification, a decidedly monotheistic belief. All particles and forces moving and acting on a given space-time are reduced to vibrations and ruminations of a single object, the eponymous string. Since one of the forces is gravity and string theory is a quantum theory, this approach must ultimately lead to a quantum theory of the gravitational force. But the structure of space-time, though indirectly brought in contact with the quantum world, is left largely untouched. As far as space-time is concerned, string theory does not go much beyond general relativity. Only loop quantum gravity treats space-time and matter on an equal, though not completely unified, footing. Here, the concepts of quantum theory are directly applied to space and time, resulting in an atomic structure analogous to that of matter. Even space-time, the stage on which matter moves, is constructed within the theory of loop quantum gravity, while much of it is presupposed in string theory (figure 10). If string theory is monotheistic, loop quantum gravity is pantheistic. Everything—space, time, and matter—is a fluctuating discrete mesh whose internal relations are what we perceive as change.

The mathematical underpinning of each theory is equally challenging and sophisticated, yet strikingly different. Both are geometric, which is characteristic of modern physics. But string theory uses continuous versions of geometry, while loop quantum gravity is based on the discrete version it rapidly developed in its first years. Both make extensive use of quantum field theory, the universal language not only of particle physics but also of condensed matter physics, but different versions of them in accordance with their distinct viewpoints on space-time: quantum field theory *on* space-time in string theory, but quantum field theory *of* space-time in loop quantum gravity. Several exactly treatable models are known in both

10. Many theories of quantum gravity describe properties of space-time by one-dimensional objects such as the strings of string theory or the loops of loop quantum gravity. But they differ considerably in their concepts as well as in the meaning of the one-dimensional objects. In string theory, strings provide excitations of matter or gravitational waves on a given space-time, as symbolized by the white rock on the right around which a thread is winding. Loop quantum gravity, by contrast, attempts to build up everything, including space-time itself, from one-dimensional loops (left)—a much more complex undertaking. (Sculpture by Gianni Caravaggio: *Catturatore di volumi* [*Volume Catcher*]), 2005. Photograph: Paolo Mussat Sartor.)

cases, theoretical highlights that always provide an elevated degree of mathematical access and sharpen concepts as well as expectations. But they all refer to different situations, and so do not provide easy comparisons of the approaches.

Most important is the difference in extra baggage between the theories, the institutionalization of beliefs and ambitions not directly serving the stated aim. Loop quantum gravity is applied directly to space-time as it is known in general relativity, just endowed with quantum laws. This combination has several new and often surprising consequences, allowing the theory to go well beyond general relativity. But these new phenomena are derived from the underlying principles and their realization; they are not by themselves added to the existing established framework. For string theory to be formulated consistently, it turns out that it requires more than just the three dimensions—height, depth, width—that we know from experience. It leads to many more kinds of particles than seen to date, and crucially rests on a new type of (super)symmetry relating different types of particles. So far, such concepts have been introduced for purely theoretical reasons. All this could be revolutionary, but it could also be wrong. The danger is that too much additional input, as is often required in specific models of string theory, can easily make the theory unpredictive if nearly everything can be encompassed in its breadth.

As for the problem of singularities and the big bang, both approaches recognize it as a crucial one and have made suggestions for potential solutions. But here, too, the vast scope of string theory does not allow for the situation to be as specific as one would like. There are very different proposals to address the singularity problem within string theory, such as the concept of our world colliding with another one embedded in a space-time of more than four dimensions (the ekpyrotic scenario developed by Justin Khoury, Burt Ovrut, Paul Steinhardt, and Neil Turok), the big bang as a collapsed but nonsingular and close-knit community of particles moving faster than light (the tachyon condensate introduced by Ashoke Sen, then used for cosmology by Eva Silverstein and Liam McAllister), or several versions of rebounding models somewhat similar to what we are going to see in loop quantum cosmology. In none of these cases, however, has a complete transition through the big bang singularity been achieved; assumptions about the transition are still necessary. This situation of a multitude of nonspecific scenarios makes it hard to see, at the current stage, how string theory might be able to deal with the singularity problem. Loop quantum gravity, on the other hand, directly addresses the structure of space-time, and does so with much greater (though still incomplete) clarity.

Before describing cosmological consequences, we must introduce more of the concepts of quantum gravity.

LOOP QUANTUM GRAVITY:
SHEDDING LIGHT ON DARKNESS

The history and development of loop quantum gravity is a fascinating one, a case study illustrating the process and perils of theoretical science. Several important contributions turned out to be incorrect, but nonetheless influenced many crucial developments. Enthusiasm was often followed quickly by sobering realizations of the limitations of what had been achieved. Faced with a vast unexplored area and fully aware of the importance of the subject, the initiators of the field demonstrated considerable courage. When applying traditional methods such as had been used for matter to a much more elementary concept such as space and time, one initially can hardly see any ground to stand on, nor any sign to show the way. Where do we start in such a situation, and where do we go from there? Every decision in these early stages is bound to have a lasting impression on the emerging field.

EARLY HISTORY: BRAIN GAMES

The complicated first five years of the theory were particularly interesting. Intellectually, the conditions for the few participating researchers were extreme. So many roads could be imagined, some ending in glorious success, most others in complete failure. Each of them would take years of calculations to follow, showing where it was leading only in the end; the first steps, the initial directions, thus had to be chosen with utmost care. Even for a theoretical physicist it is not easy to endure the corresponding double uncertainty of unknown scientific avenues combined with career insecurity.

In fact, not all theoretical physicists are created equal, and only a subclass is typically found at the very inception of a new field. The events of the first years of loop quantum gravity admirably demonstrate the types of researchers in theoretical physics. First, there are those with a physical and those with a mathematical leaning, the former primarily interested in how the world is understood, the latter in how it is described. Moreover, each of the two classes splits itself in two: the innovators, and the consolidators who substantiate what has more vaguely been conceived. One may call the poles of these double dualities the *empirical* and the *conceptual* physicist, and the *mathematical constructor* as distinct from the *analyst*.

In 1990, Carlo Rovelli and Lee Smolin conceived the idea of loops of gravity, giving the name to the field of *loop quantum gravity*.[1] The concept of loops itself was not entirely new; for a long time, it had successfully been used in quantum theories of forces such as the electromagnetic one. In those theories, loops arise not as fundamental concepts but rather as useful ways of approximation. Instead of considering a quantity, such as the magnetic field, at all points in space, one would take into account only its cumulative effect over the whole extension of a curve.[2] As when approximating a complex curve by several short straight lines, the mathematical description can simplify the picture considerably, and be more amenable to computer-aided solutions.

The idea of loops was not new, but in the hands of Rovelli and Smolin it became truly revolutionary. When used for gravity, loops are not approximations but provide a complete description; they become fundamental. The key for this realization is *general covariance*, the basic concept behind general relativity, which holds that there is no absolute position in space, but only relations between different objects matter. Changing the positions of loops has no effect when they describe gravity, while it would change the state if they are used to describe other forces. What matters is only the relationships between different loops, such as the way they may be knotted or interlinked. A complex but elegant picture arises in which space is woven from one-dimensional objects mathematically described by Rovelli and Smolin's loops.

After this realization, Rovelli and Smolin quickly derived further consequences. The area and volume of spatial regions cannot be stretched to arbitrary values, as in traditional geometry, but only assigned those in certain discrete sets. The allowed values for areas or volumes form grids, or spectra, in the set of all numbers, and only values at the nodes of the grid can be realized. In this way, the atomic nature of space is established mathematically: Regions can grow only by adding spatial atoms of specific predetermined sizes. Adding up the different atomic volumes leads to the lattice sets computed by Rovelli and Smolin. A region growing dynamically, such as part of an expanding universe, is subject to specific laws determining how the atomic interactions take place. Rovelli and Smolin also formulated an early version of these rules for changing atoms of space, which was exceedingly complicated but seemed workable.

When they started to look at possible realizations of these rules by solving the underlying mathematical equations, enthusiasm rose. While not many specific solutions for space-times in general relativity are known, Rovelli and Smolin were quickly able to find infinitely many solutions to the atomic rules they had laid down. That would indeed be exciting; it might suggest that the quantum theory of space and time, despite all expectations, could be simpler than the classical analog. Alas, this was one of the misguided yet crucial excitements that often characterize an emerging field. All the solutions they found, it turned out, described spaces of vanishing volume (but nonzero areas), none of which could possibly describe anything in the real world.

Nonetheless, with these initial strokes, Rovelli and Smolin considerably extended the previous frontiers of quantum gravity. They were driven mainly by intuition and a desire to understand the physical concepts of their theory; they represent the two types of physics-oriented theoretical researchers. Lee Smolin is a physicist's physicist: a daring explorer, not afraid or ashamed of undertaking bold raids deep into unknown territory even at the risk of coming home beat up. He is most creative in suggesting possible physical phenomena that may be new but also wrong. Carlo Rovelli is the prototype of the philosopher-physicist, who has been found citing Galileo's *Dialogo* or Newton's *Principia* and can indulge in subtleties of cutting-edge quantum gravity research. Skeptical of too-narrow mathematical procedures, he is willing to suspend the rules of known physics in hopes of finding new, more powerful ones.

Rovelli and Smolin's work was possible only thanks to a reformulation of general relativity undertaken by Abhay Ashtekar in 1986. Ashtekar is a mathematical physicist, an analyst equipped with technical brilliance and unparalleled mastery of the dark art of scientific power play. Senior to Rovelli and Smolin, he quite naturally saw the leadership role fall into his hands. And he seized it. It helped that his earlier work in mathematical relativity had endowed him with significant stature in the field (as well as personal force). Now he summoned a powerful gang of several younger postdocs to work out the required mathematics in all its details.[3] The result was a form of geometry similar to what underlies general relativity, but suitable for discrete and atomic space-times: a new type of mathematics combining that of general relativity and quantum mechanics. In just a few years a beautiful, striking framework was forged, a beacon that later attracted many researchers and students entering their research careers (including me).

The postdocs moved on to form their own projects in different coalitions. Especially important for the development of the field was the work by Thomas Thiemann, a mathematical physicist like Ashtekar, but a constructive rather than an analytic one. His specialty is not the detailed analysis of existing mathematics, but contrived constructions of new objects. He was still a student in Aachen, Germany, at the time when loops were introduced, graduating in 1993 (a student of Hans Kastrup, who would later become my Ph.D. adviser). After his initial collaboration with Ashtekar, Thiemann started to work on "transforms on spaces of connections." It is not important to explain here exactly what that means; suffice it to say that it can warm the heart only of a true mathematician. But for Thiemann, this was the way to destiny. In the process, he discovered several identities that, as he realized in 1996, could be used to make the dynamic laws formulated by Rovelli and Smolin much more solid and specific.[4] With this contribution, loop quantum gravity had achieved its characteristic, though still unfinished, shape.

It is interesting that the early years of loop quantum gravity were shaped by representatives of the four main types of theoretical physicists, realized in almost pure forms: The thrill-seeking phenomenal physicist who dares the wildest suggestions; he wanders through nature with powerful strides, and wears the label "crazy" with pride; by virtue more often mistaken than right, he must be constrained by all the other types. The philosopher-physicist who, focusing on the concepts, provides a solid footing for a physical theory, but may have a tendency to be impractical. The mathematical analyst who at his humble best advances and crystallizes the laws underlying a physical theory, but at his greedy worst may take known results and resell them refurbished as his own. And the mathematical constructor who endures long and tedious advances but can also indulge in convoluted detours on which he can lose track of the physics involved.

All these different types are needed to develop intuitive insights from the physical side and subject them to strict mathematical analysis for checks and balances. Despite the small number of researchers, loop quantum gravity's early development was complete in the sense that all four types were involved, which was extremely important for its progress. This was its great fortune; its misfortune was that all the types were independently present in different people. The rare complete physicist would embody all four types at once, making it possible to move ahead all the way from original ideas to sharply formulated results. Such a person would be the natural leader of a field, nurturing it from its inception and guiding it to maturity. When the types are all represented by different people, it is not intellectual qualities but secondary skills such as political ones or simply seniority that determine leadership.[5] One can easily imagine that this can be harmful to the development of an emerging field and its unity. Different personalities push in different directions; well-intentioned checks can too easily be dismissed as self-righteous criticism (or self-righteous criticism camouflaged as well-intentioned checks). This is a difficult situation, full of tensions that the initiators of loop quantum gravity were not always able to resolve in an optimal way.

MORE DETAILS: FENCES, FIBERS, FILIGREE

A fence is to be painted in two different colors separated by a horizontal borderline. To do so, we need to compare positions on its different boards, normally by moving horizontally from one board to the next taking the ground as the base. Now imagine a fence whose boards are infinitely long, without a bottom where they would be affixed to the base. We can still easily say what it means to move "vertically"; we just move along the boards. But how do we determine the "horizontal" direction without any orientation from the base to refer to?

In such a situation, we need an extra structure, such as a collection of nonvertical arrows as road signs telling us what we should mean by the horizontal direction. Following such an arrow we would move uniquely from one point on a board to some point on the next board, just as we were able to do by referring to the conventional horizontal direction. Only such a collection of arrows defining horizontality tells us how the different boards are related to each other; it may be called a connection.

Modern physics and mathematics is full of fences. This is not meant merely in the sense of researchers keeping competitors out of their own turf, but in an abstract sense: The fence is a *fiber bundle* whose boards are called fibers, the ground is the base manifold of the bundle, and we compare different fibers by means of a *connection*.

Moving along the horizontal direction as determined by the connection is called *parallel transport* or *holonomy.* A crucial property of general fences in mathematics is that in returning to the same board while moving horizontally along a circular fence, one need not end up at the same position on the board. (In situations where one does always end up at the same point on the fiber when moving along a circle in the base, the connection is called *flat.*) Depending on how horizontality is determined by the connection, there may be a displacement, which is called a *Wilson loop.*[6]

General relativity has a fence whose base consists of all of space-time and whose fibers are made up of all possible rotation angles (including changes of velocities, since we are talking about angles in space-time). Changing position along the fiber means that one would remain at the same space-time point but change one's orientation, just as we are forced to do when moving on a curved sphere. As this example illustrates, Wilson loops on this fence are a measure of curvature, and thus of the gravitational force. In 1986, Abhay Ashtekar built a new fence together with a special connection in general relativity, based on earlier work by Amitabha Sen. Here, the base is not space-time but only space, and the fibers are made only from spatial rotations. There is thus a reduction in the amount of information, since time is initially disregarded. But it turns out that all information about the space-times of general relativity can be recovered from the Wilson loops of Ashtekar's connection together with all areas of surfaces in space (or, more precisely, *fluxes,* which also have information about whether space is seen in a mirror).

These are the Wilson loops used by Rovelli and Smolin in their conception of loop quantum gravity. Ashtekar's work initially remained classical; there were no fluctuations and no quantum uncertainty, and Wilson loops as well as the areas of surfaces could take on arbitrarily precise values. Rovelli and Smolin realized that Ashtekar's connection can, much more easily than the traditional one used in general relativity, be described by a wave function, thus quantizing it. Here, one can directly combine the central concepts of general relativity and quantum theory: space-time and the wave function. A quantum theoretical description of the interplay of space and time now requires some kind of "wave function of sizes" rather than position as in quantum mechanics.

Not only, then, are the position and velocity of matter particles such as an electron imprecise, but so are the geometrical dimensions of the stage on which an electron moves. It is no longer possible to measure areas and curvature independently at the same time; to measure the area of a surface, we must align its boundaries with measuring rods and compare the marks. A very precise comparison requires strong resolutions, such as may be provided by a powerful microscope, to ensure that the corners of the surface indeed line up with the marks on the rod. Just as for position measurements in quantum mechanics, a microscope uses signals that must interact with the surface, making its edges move slightly. The stronger the resolution, the more rapid the movement. A measured surface seems to have fluctuating expansion or contraction rates, and expansion or contraction, as in cosmology, is related to the curvature—changing velocities due to position changes—of its surrounding space-time. Precise area measurements lead to more fluctuation and less precise knowledge of curvature, which is the uncertainty principle of loop quantum gravity. (Analogous uncertainty principles could be formulated for distances or volumes instead of areas; the latter are appealing because they turn out to lead to more elegant mathematics.)

Ashtekar's reformulation combined with Rovelli and Smolin's quantization has the flavor of unification, although not of the form seen in string theory. It is, as it were, a mathematical rather than a physical unification: Different forces are not reduced to a single principle, but the mathematical description of the forces is made uniform. The additional forces—the electromagnetic, the strong, and the weak—are in fact based on similar connections on fences. The fibers are not orientations or angles in physical space as in Ashtekar's formulation of general relativity, but angles in abstract mathematical spaces. For the theories, this difference plays hardly any role, and thus a multitude of methods and insights, gained for instance in quantum electrodynamics—the quantum theory of the electromagnetic field—can be applied also to gravity. But there are, to be sure, important special properties of relativity that required a wider advancement of mathematics before a complete quantum theory of gravity could be developed.

The initial mathematical work of the early 1990s had provided suitable quantization rules of loop quantum gravity which Ashtekar's areas and angles must obey, including their quantum mechanical uncertainty. At that time, it remained unknown whether there could be other forms of such rules, possibly leading to different quantum theories. In other words, the uniqueness question here, in contrast to string theory, was not yet clarified. Other quantization rules would imply different predictions. In principle one could select the correct ones by comparison with observations, but those do not yet exist for quantum gravity. Threatened by the possible existence of widely differing quantizations, the usefulness and predictivity of loop quantum gravity could seriously be questioned.

Later, though, around 2002, Hanno Sahlmann, at that time still a student of Thiemann's at the Max Planck Institute for Gravitational Physics in Potsdam, Germany, realized that the quantum rules of loop gravity, like those of string theory, could also be proven unique in a mathematically rigorous way. Formulating this proof in all its details took longer, but the proof was eventually published in 2005 by Sahlmann, together with Jerzy Lewandowski, Andrzej Okolow, and Thiemann, as well as independently by Christian Fleischhack. (In jest, Ashtekar called the proof the LOST result, from the authors' initials but also to indicate the apparent loss during the long years before its publication.)

Much earlier, Rovelli and Smolin had gone ahead to determine the first exciting consequences of their bold discovery. As mentioned, they were quickly led to conclude that Wilson loops represent, in an utterly and unapologetically abstract way, the creation and changing dynamics of atoms of space. Loops lead naturally to a discrete picture of spatial geometry, as generally expected from theories of quantum gravity: Spatial distances, areas, and volumes are generated when the loops build up some kind of lattice structure not only supporting but literally making space. The area of a surface is determined by the loops that intersect with it, as illustrated in figure 11. The volume of space depends on mutual intersections of loops. Loops are like atoms of space, endowing space with geometry such as length, area, and volume. This picture is entirely different from that usually referred to in general relativity: The fabric of space is not made of rubber, but woven from threads. (Applied to space-time, such a fine-structured picture was introduced by John Wheeler even before loop quantum gravity was devised, and dubbed "space-time foam.")

One can view the space of loop quantum gravity as some kind of woven structure, but this is a visualization of mathematical objects and not a direct, tangible image: Where there is no loop, there is nothing. Loops must build space in which light could propagate to show us the spatial structure. A single loop could not be seen even with the mightiest microscope, since no signal could travel through the emptiness surrounding it. At most, indirect hints based on what happens on larger scales are conceivable, as we will see in the next chapter.

A large space such as the current universe must contain numerous loops, given that each one of them contributes just a tiny value to the volume. Loops also can intersect or partially overlap, and even those not doing so are typically knotted to an abstruse fabric as in figure 12. Seen from afar, this weave can indeed appear like a continuous space, as assumed in general relativity.

To see how these discrete spaces appear in an actual dynamic process of change, not adding or removing loops by hand but having them interact to produce new ones or possibly self-destruct, one needs to use the atomic evolution equations. Such equations and their solutions must, first of all, explain why the individualistic attitude of space, broken into tiniest chunks, gives rise to the familiar smooth classical rubber-band structure on the larger scales we have so far probed by observations. How does this wide, pervasive fabric we have come to call space-time "emerge" from the local tête-à-tête interplay of elementary quanta? With the work by Rovelli, Smolin,

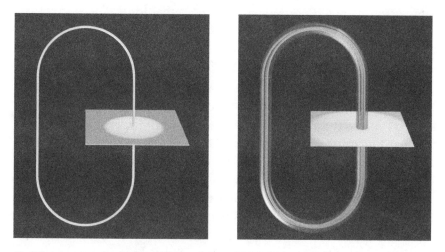

11. A loop generates area when it intersects a surface (left). For several overlapping loops, the induced area is increased proportionally (right).

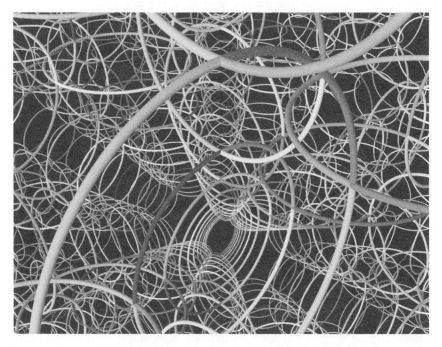

12. View of a sample fabric determining the quantum geometrical excitations of space-time by its links. The denser it is, the more continuous space appears to be. The gray scale indicates excitation levels of loops.

Ashtekar, and their collaborators at hand, loop quantum gravity was in an excellent position to address such questions.

Deriving laws for the temporal change of a quantum gravitational universe was not supposed to have taken that long. The electromagnetic force in its quantum form, with photons as the elementary building blocks, was essentially understood in the 1930s. All other fundamental forces except gravity followed with a coherent mathematical description soon after they were discovered in particle interactions. Gravity had been known for centuries and should duly receive a quantum treatment, too. Luminaries such as Paul Dirac, Richard Feynman, and others tried their best but failed. In the 1960s, John Wheeler and Bryce DeWitt started their own dedicated program, but here also there was no clear road toward its rules for temporal change in sight, despite renewed efforts in the 1980s by Stephen Hawking and others. What was missing in all these attempts was a proper understanding of the fundamental nature, down to its atomic scales, of space and time. This is what loop quantum gravity has provided.

Formulating suitable equations in a mathematically useful form, as it was realized in the mid-1990s by loop quantum gravity, surely represented important progress. When Thiemann introduced his equations, the mood was indeed extremely optimistic, almost euphoric. In principle, these equations could be used to determine the structure of space-time in its finest details. Analyzing the past of an atomic space-time, one could calculate backward in time from any given arrangement of loops to find where it would have come from. One could try to understand the universe at very early times, perhaps in a state without loops such as might have existed at the big bang singularity. Or maybe the equations of loop quantum gravity could show deviations from those of general relativity at such extreme energy densities of the early universe, possibly preventing the state of vanishing volume and thus the singularity altogether. With sufficient control over the equations, there would be no limit to the number of intriguing questions one could address. Shortly after Thiemann's feat, Lewandowski organized a conference at the Banach Institute in Warsaw, Poland, where, according to lore, the prevailing opinion was that one could now solve all the fundamental questions of the cosmos, for its ruling laws were known. (However, I cannot report this from direct experience: I was still studying at that time and did not attend the conference.) Alas, the equations' complexity quickly struck down all hopes, and to date not a single exact, or just approximate, solution able to describe an entire universe is known.

Such difficulties should not come as a surprise if one considers the immensity of the task. Posing and then solving atomic equations for the dynamic changes of the loops of quantum gravity is an enormously complex procedure, at a mathematical as well as conceptual level. Among other things, it illustrates the special and confusing role of time in fundamental modeling. There is no space-time to start with; even in the classical theory of general relativity, space-time and its structure is a derived concept that follows after solving Einstein's equation. In quantum gravity, the task is to find the analogous equation, successfully taking into account the tiniest atomic details. When solving quantum gravity, what we classically understand as space-time arises as a collapsed version of infinitely many spaces with different sizes and geometries, all in a quantum superposition. According to the rules of quantum theory, they all exist "at once," but time is yet to emerge. One can think of this superposition as a dissection of space-time, like a deck of cards, into its spatial configurations at constant times, yet unordered owing to the absence of time.

Once such a mathematical solution has been achieved, one can analyze it further. At this stage, all the spatial configurations may be ordered, for instance by arranging them by their volumes. When this is done, evolution comes in by computing other quantities for the ordered configurations, such as the matter density. One could, for instance, derive a functional relationship between the density and the volume of space quite similar to the solutions one obtains for Einstein's equation in a cosmological context. Here, however, the intermediate stages of atomic change and quantum superposition have brought in crucial aspects of quantum gravity, with an emphasis on what is happening especially in regions of high densities. The concepts of evolution and time emerged at a secondary stage; they were not present in the initial solutions.

Not just solving these equations but even formulating them is difficult. For consistency, results obtained from the solutions must not depend on what time is used for the ordering. It may be natural to use the volume for arranging the spatial configurations of an expanding universe, but that is not the only possibility. Instead, we might use the temperature, or any other physical parameter with a sufficiently strong trend. If what we derive depended significantly on what quantity we choose to order events, no predictions could be made. This leads to an important principle, known as *covariance* in general relativity; it must also be obeyed by the atomic equations of quantum gravity. While no fully covariant formulation of loop quantum gravity has yet been achieved, Thiemann's important work had provided some early hints as to how this may eventually be done.

In addition to the stubborn consistency issue, the equations are not uniquely specified; their final form remains unclear. Several of their general properties are known, based on the unique quantization rules for areas and curvature, which distinguish this type of equation from others in gravitational physics. But some of the terms in the dynamic equations are still affected by undetermined parameters. In contrast to string theory, the present status of knowledge does not at all indicate a unique dynamics even though the underlying quantum rules are unique. There is no choice as to what discrete values the steps on the ladder of volume must take; the growing universe can climb up on it in many different ways. Still, loop quantum gravity has a major advantage: With it one can unravel some properties of the universe at the big bang, perhaps not with exact solutions but at least with approximations and model systems. Thiemann's original construction, though often modified since then[7] — an ongoing process that indicates the arduous development of quantum gravity — plays a decisive role in applications of the theory.

DISCRETE UNIVERSE: WEAVING THE STORY OF THE WORLD

Loops build space in a dense mesh. Even empty space is full of them; a count in just one cubic meter would result in a number with about a hundred decimal places. This mesh is ever-shifting, for the change of loops is what encodes time. Moreover, this is a fuzzy mesh, subject to quantum fluctuations. On a fundamental level, space is rather agitated. One might visualize it as a mass of fluttering heated air above a dark road exposed to the glaring sun of a hot summer day.

Space is commonly thought of as the stage on which matter moves or signals propagate. Now, all that does happen on a mesh of loops, but nothing jumps from loop to loop (for there is nothing between the loops). Rather, as the ever-fluttering loops change their connectivity, they carry matter or signals with them, moving them ahead one spatial atom at a time; everything is a-changing. In this way, matter can, as it were, diffuse through the mesh of loops and move around. When we move an arm, all its atoms or even elementary particles must move in concert. Similarly, when an elementary particle moves, all its spatial atoms flow in concert. Even though performing the resulting motions may seem trivial to us, describing them in exact mathematical detail is in both cases an enormously complex task.

In cosmology, the stage of change is the whole universe, whose volume in loop quantum gravity is subject to the same quantum theoretical uncertainty as is known for matter. On average, it expands uniformly; but if one could view this expansion at close range, one would notice tiny variations. In addition, quantum jumps occur just as they do in the energies of atoms and molecules: The universe changes its size not continuously but in the smallest jumps—it grows brick by brick. Only because the building blocks are very small, about the size of the Planck length, does the growth appear continuous.

All of space can be seen to arise by adding loop upon loop. In this process one has to start from some initial state, or one is led back to it by performing the game backward and removing loop after loop. In the end, only an object without a single loop can remain: the State of Hell encountered earlier. Here we can see for the first time how loop quantum gravity could give insights about the big bang, for the big bang singularity in general relativity is also a state without volume (but with an infinitely high energy density of matter). To see what is going on there, it is not sufficient to remove one loop at a time by hand; instead, one must analyze how the dynamics of quantized space-time can bring this about. As general relativity describes the expansion of the universe by a space-time that solves Einstein's equations, such as the differential equations illustrated in figure 4, loop quantum gravity has equations of discrete nature describing the dynamic appearance and removal of loops. Starting with an initial configuration and its quantum theoretical wave function, the further development in time as a consequence of successive additions and removals of loops is determined.

At a precise level, the picture of a jumpy universe, structured on the smallest scales by a discrete weave rather than a smooth space-time, replaces that of a rubber band as used in general relativity. The smallest volume changes are given in terms of the Planck length and thus unimaginably tiny. A single quantum jump is imperceptible in the behavior of a large universe as we see it now. At small extensions as realized in the early universe, however, the woven texture is particularly important. Here the universe itself is so small that a continuous change of volume differs greatly from a quantum jump. The universe's expansion at those times is entirely unlike what was expected in general relativity: Quantum corrections to Einstein's equations become necessary.

As a consequence of the continuous picture, the singularity problem arose; now, quantum gravity shows strong deviations from the classical theory just in this exact respect. Could this mean a solution to the singularity problem? Can loop quantum gravity save a universe from a singularity, as quantum mechanics stabilizes hydrogen and other atoms?

LOOP QUANTUM COSMOLOGY:
ATOMS SMEARED OUT

An application of loop quantum gravity to the singularity issue, as it appears for instance in cosmology, at first seems problematic. After all, the dynamics of the loops is so complicated that not a single exact solution has been found yet, and nobody seriously expects this to change in the near future. Also, a mathematical analysis of the equations without knowing explicit solutions would have to overcome severe difficulties. But this problem does not arise only in quantum gravity. Einstein's equations of general relativity are not easy to solve either, yet we know some of their general properties. The most important example is the singularity problem, according to which all solutions, not just those few known explicitly, must, under some realistic conditions—for instance on the form of matter—reach a singularity or have come from one.

Exact solutions to a complicated theory such as general relativity can be found only in special, often highly symmetrical cases. In cosmology, homogeneity and isotropy are usually assumed: Disregarding details such as planets, stars, or even whole galaxies, the universe on a very large scale does not show position-dependent properties. None of our little fits and fights, on small contested planets or between entire merging galaxies, plays any role for the whole universe. Thanks to wide-ranging galaxy maps such as the Sloan Digital Sky Survey (SDSS) described later, this long-assumed "cosmological principle" has by now found impressive support through observations. Similarly, the universe at large appears uniform independently of the spatial direction of sight. Thus, the total expansion of the universe can well be described by simplified solutions in which only the temporal change of its volume is taken into account. Abandoning many details implies an immense simplification of the equations:

Just a single function is to be considered instead of many whose behaviors would strongly depend on one another. In such cases, plenty of solutions are known and are being used to evaluate cosmological observations. That the restriction to symmetrical solutions is not too strong even for conceptual questions is shown by the main issue of interest here: Under this simplification, the singularity problem still arises. The volume of a homogeneous and isotropic universe vanishes at some time, and the theory breaks down.

Loop quantum gravity is even more complicated than general relativity; one should not expect that simplifying assumptions can be dropped. But this is no reason to give up. If one could describe at least the highly symmetrical examples in quantum gravity, an investigation of the singularity problem would be possible; even these simple classical solutions challenge general relativity sometimes. An analysis of space-times in quantum gravity satisfying the cosmological principle, a subset called quantum cosmology, is already worthwhile. How can such a simplification be performed? It looks obvious if symmetry is demanded for a smooth rubber band such as general relativity's space-time. But what remains of the loops of quantum gravity? A symmetrical weave should become more orderly, as in figure 13 compared to figure 12, but even such a regular lattice is not homogeneous: Its links differ from the nothingness in between. A spatial weave woven from its links would be completely blurred by symmetry requirements as strong as homogeneity. One could worry that no trace of the lattice properties of quantum gravity, including the quantum jumps of volume indicating crucial effects in a small universe, would be left.

Fortunately, and perhaps surprisingly, it turns out that the most important properties do in fact continue to exist even if homogeneity and isotropy of space are imposed. Such a symmetrical situation does not allow single loops in a space made uniform, but the change in time, especially of the volume, still occurs in smallest jumps. There is, moreover, a loopless State of Hell in which the volume vanishes, much as at the big bang singularity. In 2000 and the years following, the mathematical methods developed in the 1990s allowed me not only to formulate such isotropic geometries in the newly minted loop quantum cosmology, but also to analyze their temporal evolution as defined by Thiemann.

An overarching quantum theoretical investigation of the singular-

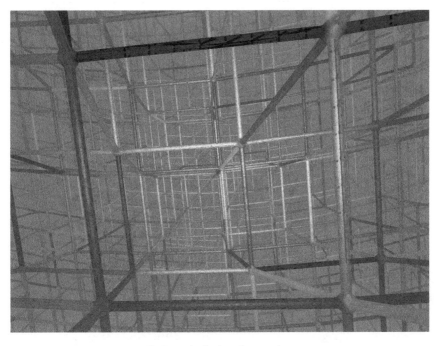

13. A regular lattice of space-time.

ity problem thus came within reach. Still, the concrete application did not arise directly, for even the equations with the highest possible symmetry initially appeared too complicated. The breakthrough happened with a little help from chance, as I can directly report from my own experience, in an example that illustrates how external influences can act on scientific developments. Progress does not happen as logically as it may sometimes seem from the outside.

After I had imposed the first set of symmetry assumptions to simplify the equations, they remained too intractable for an analysis of quantum cosmological evolution. A further possibility of simplifications did suggest itself, but seemed to me to be mathematically invalid—for good reasons, I thought at that time. This point was a decisive one, for the equations without concrete applications would be of little value. Realizing homogeneity and isotropy in spite of the not at all symmetrical form of loops may have posed a mathematically interesting issue, but on its own it does not show the way to new physical phenomena. There was an additional difficulty: Since the 1960s a theory of quantum cosmology has already existed— founded by John Wheeler and Bryce DeWitt as a spinoff of their attempts to quantize gravity, and further developed by physicists such as Charles Misner as well as, somewhat later, Alex Vilenkin, Jim Hartle, and especially Stephen Hawking. Claus Kiefer, among others, has also contributed a series of detailed studies to the understanding of the semiclassical behavior, a fundamental issue now important for an explanation of the emergence of structure in the early universe.

All this happened before the introduction of Ashtekar's formulation of general relativity and of loop quantum gravity by Rovelli and Smolin, and so this form of quantum cosmology was not based on discrete space. How, precisely, space on small scales looked in this theory was largely unknown: The theory could reliably be formulated only for the homogeneous spaces of cosmology, but not more generally. At the end of the twentieth century, the prevailing belief among researchers, aware of the old quantum cosmology as well as of loop quantum gravity, was that these are two sides of a single theory: For solutions without any symmetry, one would have loop quantum gravity with its complicated construction of space from discrete building blocks. But when space is washed out by enforcing symmetry, affording a view of it only on scales much larger than the extension of spatial atoms, one would obtain the old quantum cosmology. A further analysis of the symmetrized equations of loop quantum gravity would have been futile, for it could not have resulted in anything new. Indeed, leading figures in loop quantum gravity at that time had repeatedly advised me against wasting my time by further following the symmetrization.

A direct reproduction of the old quantum cosmology as the symmetrical special case of loop quantum gravity would, in the context of the singularity problem, be devastating. If the old theory were the unavoidable description of cosmological space-times in quantum gravity, singularities could not be removed, for this problem remained unresolved: As a possible quantum theory of the universe, quantum cosmology as developed by Wheeler and DeWitt does lead to a wave function and quantum theoretical uncertainty. But its washed-out wave packets simply run into a singularity comparable with the classical one. There is still a point in time when—disregarding the unavoidable uncertainty—the volume of the universe vanishes and its temperature diverges. Nor can the quantum cosmological equations of old, like Einstein's, withstand this situation:

They lose their mathematical validity, keeping shrouded what happened at the big bang.

In retrospect, the explanation of this failure is simple: The initial version of quantum cosmology introduces some quantum effects, but it hesitates before the decisive step. It overlooks the discrete nature of space and time, as later demonstrated by loop quantum gravity. Near the singularity, the behavior changes slightly compared to general relativity, but not crucially so. Only loop quantum cosmology was able to highlight this fact, and to provide a concrete physical mechanism for preventing singularities.

To see all this, one must first solve the equations of loop quantum cosmology, a seemingly difficult task. And yet it did turn out to be possible; this is where chance entered the game. After I had constructed the symmetrical formulation of loop quantum gravity, or loop quantum cosmology, for my Ph.D. degree with Hans Kastrup at RWTH Aachen, Germany, I was for some time distracted from its equations. In 2000, I was moving to Pennsylvania State University, where Ashtekar had offered me a postdoc position. (He did so even though he was then among those not convinced of the promise of loop quantum cosmology. After a great deal of explaining and persuading on my part,[8] he is now vigorously using loop quantum cosmology in his own research and has contributed to several recent developments in the field.)

After the move, I had to return to work, but was no longer familiar with all the details. In particular, I had briefly forgotten the reason for the apparent invalidity of the crucial simplification. It all happened during one of the hot, humid late summers typical of the eastern United States, which I had not yet grown used to. I still remember the numbing daytime heat and cricket-song nights of that time—as well as the correspondingly weakened concentration combined with a lowered level of inhibition that tends to make one forget mathematical worries. I exercised the simplification, solved some of the equations, and saw the loop universe before its big bang.

But was this allowed? The calculation's promise, including a possible resolution of the singularity problem—a major issue that had plagued gravitational research for decades—was too enticing to drop everything again. After the first excitement I remembered why I had not undertaken the simplification much earlier. There was, after all, its apparent inconsistency. With a fresh view of the old problem, I was fortunately soon able to show that the crucial simplification was not only allowed mathematically, but even necessary for implementing the symmetry completely. My initial version, without the final step, would have left space slightly squashed and distorted. With this last bit of asymmetry removed, the equations easily became amenable to standard treatments, producing explicit solutions and showing general properties.

In this way, the first results of loop quantum cosmology, developed much further in the ensuing years to lead to the universe scenarios described in the remainder of this chapter, were put on a solid footing. Now even space-times not strictly obeying the cosmological principle can be tackled, an extension of the original calculation to which many researchers have started to contribute now that its promise has been shown. Beyond the immediate implications for kick-starting loop quantum cosmology, this experience illustrates how strongly stubborn opinions can influence scientific work, or even handicap it—and what progress can come when one occasionally steps aside.

LESS IS MORE: GAINING TIME

> There is nothing more precious than time, for this
> is the price of eternity.
>
> **—LOUIS BOURDALOUE**

The universe resembles a long-distance runner, truly the hardiest ever seen. As every runner knows, one occasionally approaches physical breakdown. Muscles weaken, motion becomes uncontrolled. While a run at full strength appears to be a single smooth and sleek process, safely getting through weak periods requires every single step to be precisely measured. Too large a step costs yet more strength; too-timid ones release all muscle tension and lead to stumbles.

Right now, the universe boasts immense strength; to mock us, it even seems to have initiated a spurt a while ago—an acceleration of its expansion to be encountered in the next chapter. (It seems to have been fired up by the spectators: As can be seen from the escape velocities of distant star explosions, the acceleration began, compared to other cosmological time scales, relatively recently, "just" before mankind started its cosmological observations.)

The big bang, by contrast, resembled a period of weakness. Here, the universe was heating up and came close to a breakdown, a trouble reflected in the existing theoretical descriptions of this phase: General relativity still assumes a continuous run but quickly breaks down in a singularity. Quantum theories of gravity are more careful and pay more attention to the precise step size. Is this enough to survive the big bang without damage?

Loop quantum cosmology has brought exactly this consequence. First, by the interactive process of changing loops of space it introduces a discrete time, thus improving the old version of quantum cosmology. Time cannot change arbitrarily, but only in multiples of a smallest time step: Change happens whenever the meshed loops reconnect, but not in between these discrete stitches. Loop quantum cosmology grants the universe less time, as it were, since all moments of time falling between the grid's smallest steps are eliminated. This happens on microscopic scales and is, due to the smallness of the step—again, of Planck size—largely imperceptible to us. But much more impressive, combined with this microscopic *less* of time is a macroscopic *more* of time. The timeline in the long-distance domain of the entire universe is extended: Time does not end at the big bang.

There is a prehistory of the universe before the big bang, with space and time as they cannot be seen in general relativity. As Dirac's combination of special relativity and quantum theory had led to a new world of antimatter, a combination of general relativity and quantum theory provides a new world of space and time. This universe before the big bang is shrouded by a veil, penetrated not by the equations of general relativity but by those of loop quantum cosmology. Before the big bang, the universe was shrinking; it collapsed to smaller sizes under its own weight, eventually giving rise to the hot dense phase of the big bang. The transition itself can be analyzed only with quantum theories of gravity. Limits of the classical theory are transcended by a more encompassing theory, one that takes into account quantum theoretical properties. The big bang singularity turns out to be a limit of the language in which it was first formulated; it is not a limit of the world.

How is this possible? How can quantum physics prevent the universal collapse to a single point and the corresponding unbounded rise of temperature? How can gravity, after it had once recklessly caused the collapse of its own embodiment, space-time, repent? As in the stability problem of atoms, solved by quantum mechanics, the big bang singularity is prevented by new repulsive forces in loop quantum cosmology, counteracting the collapse caused by the purely attractive classical gravity. The collapse of a contracting universe, when it is small enough to bring quantum theory noticeably into the game, is first slowed down, completely stopped after some time, and then turned into expansion. One can roughly imagine a part of our universe's history thus: Before the big bang it was contracting, opposite from the currently realized expansion. By collapsing, it became ever smaller and hotter, entering the big bang phase. Quantum effects were then prevailing and the slowdown as well as the rebound into the now visible expanding universe came about.

Details of this history, such as precise properties of the universe before the big bang, are much more difficult to come by. Did it look like our known expanding branch, except for the reversal of expansion? Did the reversed expansion have further consequences, such as a turnaround of one's perception of time, causing one to remember the future and predict the past? Was it, anyway, so classical in its space-time structure long before the big bang, even at the large volumes it might possibly have had, that we can compare it with the present form of space-time? Or were quantum effects perhaps predominant even for space and time, such that the volume, though contracting on average, was subject to stronger fluctuations and jumpiness? Did it allow conditions that would make life possible? And where did this collapsing universe come from, and where is our expanding branch heading? Will the current expansion one day turn into a collapse, such that we will, like our preceding universe, move toward another big bang for the next world? Did the universe before the big bang come from such a recollapse of an older universe, one that rose from an ancestral grand big bang? Is there an infinite succession of such big bangs and intermittent expanding and contracting phases of the universe?

To answer these questions, one must enter much more deeply into the theory. Most of them cannot yet be solved reliably, and many such questions will probably never be reasonably treated by physical theories. Rather, they border on philosophy. What is known so far will be related in the course of this book.

THE PREVENTED SINGULARITY: REPELLING A WHOLE UNIVERSE

In the mathematical language of loop quantum gravity, a universe collapsing as a result of overpowering gravitational attraction, a world shrinking through the disappearance and fading away of its atoms of space, approaches the State of Hell, the reign of absolute shallowness of space. The history of the universe becomes one of utter turmoil, chaos, and confusion. Space is fluctuating and disintegrating, much worse than any shaking ground. All forces struggle for dominance, raining punishing blows not just on matter but even on space and time. In this climactic battle, an ancient power is rising to absolute prominence. Gravity, belittled by its fundamental competitors when tempers were still low, now holds the key. It sets itself against the wide converging tide it had released, then calms the seas and lets the universe expand to vent its anger.

How can quantum gravity concretely provide forces counteracting its own classical attraction? Mathematically as well as intuitively, this is a direct consequence of discrete time: Taking all points in time together, there is not a continuous line but a lattice, or grid. Such a time lattice has only a finite amount of storage space for energy: Like a sponge, which can absorb a limited amount of water but when fully soaked repels surplus water, the time lattice acts repulsively once too much energy threatens to be present at one time.

Here, the quantum nature of all matter is important, requiring wave functions to be used for its description. Energy is encoded in the frequency of the wave function, that is, the number of oscillations per unit time interval. As with light, the higher the frequency, the more energetic the wave. Ultraviolet light of high frequency is much more energetic and damaging than the lower-frequency infrared light. High frequency, on the other hand, means that the oscillation length of such a wave is small: Only a brief amount of time passes between two successive intensity maxima of the wave. A continuous time line could support very short oscillations, and thus any amount of energy. A time grid, by contrast, can allow only oscillation lengths larger than its grid size. Such a wave would become discrete and distorted when drawn on a lattice, but it would still be a recognizable wave. Shorter waves, however, would simply disappear, since their oscillations would fall between the gridpoints. In this way, a discrete time cuts off too-rapid oscillations, and too-large energies (figure 14).

Owing to the smallness of time steps, the lattice can absorb a great deal of energy, but not an infinite amount. While not essential in most regimes, the resulting energy bound is important at the big bang, the most highly energetic event in the universe. According to general relativity, energy densities in a contracting universe should grow unboundedly, an expectation incompatible with a lattice of finite storage space. A consistent theory with a time lattice, as it arises in loop quantum cosmology, must cause repulsion of surplus energy. But there is nothing outside the universe into which energy

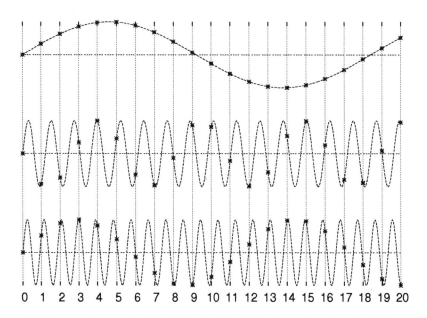

14. If we sample waves by taking their elevations only at the nodes of a grid with fixed spacing, for instance as the result of the atomic nature of space, short wavelengths are suppressed. The top wave with a wavelength larger than the spacing is not distorted much by the sampling; its smooth behavior is still recognizable by the discrete set of values it takes at the grid points. When the wavelength comes close to or is smaller than the spacing, however, the discrete samples give a picture very different from the original smooth wave (bottom two waves). In particular, oscillations of the sample points show a wavelength always larger than the grid spacing, even if the original wave had a smaller wavelength.

could be pushed. Too large an energy can be prevented only by stopping the collapse itself, the cause of the energy growth, and turning it into expansion. In this way, a *less* of local times in the lattice provides a physical counterforce to the collapse, and thus a *more* of time beyond the big bang.

For a large universe of low energy density the discreteness is unimportant, but it is essential at small size and high energy. Concerning the discreteness in time, a universe looks like an hourglass. At the wide top part, far from the narrow opening in the middle, the sand appears to be sinking down continuously. The sand's discreteness, composed of individual grains, does not play a role here. But looking at the narrow opening in the middle, one notices the graininess of the contents. Time is not measured continuously but discretely — one grain at a time. Below, the glass opens up again and the sand seems to rise without noticeable discreteness.

In contrast to quantum cosmology, whose time is passing discretely at a fundamental level, an hourglass only *measures* time discretely. Choose more finely grained sand, and time would be measured ever more continuously. The graininess is naturally limited by the atomic structure of matter, but other methods, such as modern atomic clocks, allow much more precise time measurements than are possible with the falling time of sand grains through an hourglass's opening. With these technologies, one can further subdivide the time steps. But at the level of the quantum theory of spacetime, all attempts to divide units of time any further reach a final limit. There is literally nothing between two successive atomic points of time in the discrete lattice of quantum gravity.

At the level of a discrete structure of time, the prevented big bang singularity is connected with a further phenomenon: Time before the big bang implies a reversal of spatial orientation. It is as if the history of the universe before the big bang would, viewed from the present, have happened in a mirror. A right-handed person who survived the big bang would thereafter be left-handed. Space is inverted on itself, like a sphere turned inside out. Here, the collapsing universe resembles the surface of a deflating balloon. After some time, the rubber would collapse airlessly, ending the process just as the classical collapse of a universe would end in a singularity. But if we assume that the rubber pieces could somehow penetrate each other in an uninhibited way, the balloon would reinflate after the collapse.

If, moreover, the rubber simply follows its initial motion by inertia, the former inside must now point outward: Orientation reverses. In the same way, spatial atoms penetrate one another at the big bang and invert space.

How is it possible that the universe avoids total collapse into the nothingness of space, and yet inverts itself? Would it not be required for space to shrink to a point at the moment of inversion, or at least be squashed down to vanishing volume, if the collapse happens anisotropically? Also here, the discreteness of time, combined with other key properties presented by quantum theory, is crucial. The balloon does not deflate continuously but in a jumpy way, much like a movie made with a finite number of frames. It turns out that the elementary dynamics of loop quantum cosmology, which already gave rise to the discreteness in the form of having only finitely many frames, automatically takes out the one that would correspond to a completely collapsed balloon. Even if one tries to arrange the time steps in such a way that time zero, when the full collapse is reached, would be included as a frame, that frame is automatically detached from the rest and does not cause a singularity. The resulting movie shows a deflating balloon, that never collapses completely and yet inverts itself and reinflates.

Finally, in addition to discrete time, there is a further mechanism causing a repulsive force—a kind of double insurance by quantum theory against total collapse into a singularity. Not only does the space-time lattice show clear deviations from the known macroscopic behavior at small distances, but matter energies in such realms also behave differently than expected classically. In this context, it is useful to remember the behavior of black body radiation, as explained by Planck: According to classical expectations, its energy density should grow without bounds at small distances, presenting an example of a singularity problem.

As computed by Planck, and confirmed by direct measurements, this increase does not happen. Quantum theory implies that energy is distributed not continuously but over small, discrete packages, the photons of the radiation. As before, small wavelengths—where classical physics would make us believe that the energy density of black body radiation diverges—correspond to large frequencies, or to large energy packages. Since the total amount of energy in the radiation is fixed, it cannot fill the large packages required at small wavelengths. For small wavelengths, fewer packages are filled, and the energy density starts to decrease; for the smallest ones, no energy is present at all. Thus no divergence arises.

Now consider matter in a collapsing universe. With the whole universe shrinking, all scales of distance as well as the wavelike matter it contains will eventually be small. Normally, one would expect densities—obtained by dividing the mass of matter by the volume of a region—to increase when distance scales decrease. But as in the quantum theory of black body radiation, loop quantum cosmology shows that instead, in a sufficiently small universe, less matter energy is distributed over all atomic space: Quantum waves do not find enough room. As in Planck's description of heat radiation in a box, the classically expected increase of densities all the way to infinity is interrupted by quantum theory, and instead turns to a decrease toward zero.

Now the problem is less threatening. But if the universe keeps collapsing, its equations could still break down even with finite matter densities. However, repulsive forces counteracting the collapse arise. Matter energy, after all, determines the form of space-time and its curvature, in turn responsible for the gravitational force. While the classical form of energy always causes attractive gravity, the turnaround at small lengths, as it happens in a dense, tiny universe, can lead to a redirection of the force, an additional, independent mechanism for a repulsive force provided by loop quantum cosmology. (Most big bang scenarios in string theory are based on such matter-related effects, without specific insights into the structure of space-time.)

Such counterforces do not seem to be a random result, but are a general phenomenon in this kind of quantum gravity. They do not just appear in some special cases—for instance in detailed solutions analyzed numerically in a set of stimulating articles by Ashtekar, Tomasz Pawlowski,[9] and Parampreet Singh in 2006—such properties of quantum cosmological space-times can be found in many situations. Still, the phenomena have not yet been investigated completely. And even though they will reappear in our later description of the collapse of black holes, it has not been proven that sufficiently strong counterforces, powerful enough to prevail against any kind of singularity, are a general property of loop quantum gravity. Here, research is still ongoing.

ON THE USE OF MATHEMATICS—AN ILLUSTRATION:
CLIMBING THROUGH THE STATE OF HELL

Loop quantum gravity determines states for three-dimensional spaces of different sizes by wave functions describing the usual geometry of areas and volumes in a quantum and fluctuating way. Not all spatial sizes are possible; rather, there is only a discrete set, just as for energies in atomic spectra. When the spatial weave of loops is extended by a new knot, the total volume changes by a fixed amount of a specific value.

In full generality, the allowed values for volume, assumed by such a spatial foam fluctuating on the smallest scales, are difficult to compute. One can see this in figure 15, which shows an example of values of the volume in dependence of a parameter controlling the asymmetry. Although much headway has recently been made by the calculations of Johannes Brunnemann and David Rideout, no complete picture is known for the general case. Even spatial structures that appear uniform on large cosmic scales are fundamentally constructed in an atomic way, just as a material body consists of the well-known atoms. To construct a macroscopic object—be it a crystal or just a piece of empty space—from single atoms in a mathematical description, keeping exact control over its properties such as energy or volume, is an extremely complicated enterprise.

Luckily, complete control is not always required. When we are interested only in, say, the large-scale behavior of cosmic expansion, crucial simplifications result. Once they are realized, the spectrum of all possible volumes becomes calculable concretely and in detail. Moreover, for every value of the volume there are only a few states—two, as it turns out (except for the vanishing volume of the singularity, granted a single unique state).

When I first investigated these relevant structures in loop quantum cosmology, I was puzzled by this doubling. But there is a clear and far-reaching explanation: A state is characterized not only by volume, but also by orientation; space and its inside-out replica are distinct from each other. In two dimensions, one can visualize this as a balloon, which can be blown up to a certain size in two ways: with or without being initially upended, turning its inside out. This is in

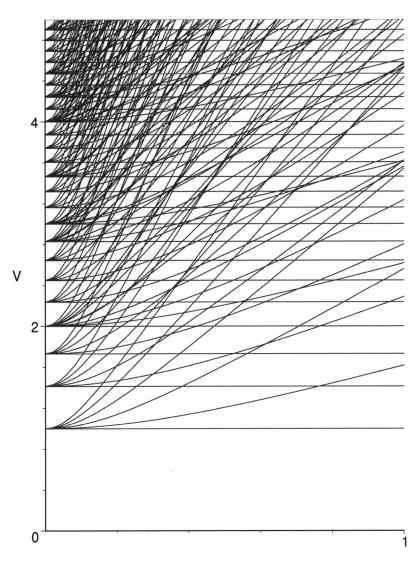

15. The ladder of volume, changing when asymmetry is continuously switched on along the horizontal direction. On the left, the ladder is regular and easily computable; on the right, it looks more and more messy. The behavior of this level splitting is analogous to an effect in the energy spectra of atoms, whose levels change and split if new forces such as a magnetic one are switched on, breaking initially realized symmetries.

agreement with the fact that vanishing volume has a unique state, for a space devoid of extension has no inside to be turned out. Mathematically, this degree of freedom is called the orientation of space.

Even more complicated than the allowed volume values is their behavior in time, the dynamics: With the volume values we know the ladder a universe can grow on, but how precisely does it climb up? How fast must it expand, and can this be accelerated? Or, winding back time to approach the big bang singularity, how does the universe climb down to small sizes into its potential demise? Reaching the unique state of vanishing volume, do we touch down hard on the singular ground, as if using too short a ladder, which, if we lose the grip of the lowest rung, makes us fall abruptly into the abyss?

After I had analyzed the doubling of states due to orientation, I had already noticed its possible meaning for the dynamics of a universe. Instead of considering the set of all volumes as a twin pair of two unidirectional positive axes, each starting at the singularity, we can arrange them as a single axis made of positive and negative numbers, the singular zero now lying in the middle rather than at a boundary. Orientation is indeed described by a sign factor, determining whether a state is counted positively or negatively. We have not two separate ladders, but a single one stretching right through the singularity.

It quickly turned out that the temporal evolution of the universe does indeed take place on this long ladder. With the symmetrical conditions of the large-scale universe, I finally managed to simplify the dynamic equation—provided by Thiemann for general states—sufficiently to find solutions. It has a compact form, of a *difference equation*, by now used in numerous scientific works:

$$C_+ \, \psi_{n+1} + C_0 \, \psi_n + C_- \, \psi_{n-1} = \hat{H} \, \psi_n$$

Here, ψ_n symbolizes the state of the universe—the wave function—at different values n representing the ladder rungs: the volume values, together with a sign for orientation to indicate the part of the ladder we are on. Moreover, there are coefficients C_+, C_0, and C_-, whose form ensures the correct correspondence to Einstein's equation, and $\hat{H}$ describes the matter content of the universe. The equation connects states at different sizes with each other, telling us about the quantum mechanical growth of the universe: One can successively solve the equation for ψ_{n+1}, the value at the $n+1$st rung, by relating it to the values at the preceding rungs n and $n-1$. Allowing n to run through all integer numbers, one obtains the wave function for any number n, if one only specifies initial values at two chosen rungs. As general solutions of the equation show, the growth does

16. Two sides of space, which, in many regards, behave like mirror images or inversions of each other. (*Universo positivo* [*Positive Universe*], 2002–06; *Universo negativo* [*Negative Universe*], 2002–04. Sculptures and photograph: Gianni Caravaggio.)

not cease when n assumes the value zero of the singularity. What comes to an end at the singularity are only the classical equations of gravity; quantum cosmology lives on. Here, discrete space can turn itself inside out without damage and continue to crawl on its curve.

SO, WHAT ACTUALLY HAPPENED BEFORE THE BIG BANG?
THE ULTIMATE QUESTION

No direct observation of the universe before the big bang is possible. In the extremely compact state near its turning point, the early universe is simply too opaque for light or other forms of electromagnetic radiation to reach us. The earliest moment visible in this way was after about 380,000 years of expansion. At that time, the universe had grown enough to cool down and dilute matter, making space transparent. In a universe still at very high temperatures, about 4,000 degrees Celsius, matter radiated like a hot, glowing body akin to the surface of a star. Before that time, matter in the universe was even denser and hotter and radiated in the same way. But the radiation was directly reabsorbed into the fiery plasma of the early universe. Just as one can see only the surface of the sun and not the interior, we cannot glance back arbitrarily deeply into the big bang.

IMPENETRABILITY: THROUGH THE FOG OF OLD

Only after sufficient dilution and cooling of the still hot matter was it possible for some of the radiation to escape and start traveling through the expanding universe to us. As drops of rain condense in a dense fog and slowly improve the view, below a temperature of 4,000 degrees Celsius, neutral atoms formed in the early universe. Without a net electric charge, atoms scatter light much more weakly than unbounded electrons and protons. Traces of the radiation released after sufficient expansion can be observed with sensitive antennae as cosmic background radiation. They are one of the most important sources of information about the early universe, as we will encounter in more detail in the next chapter.

All this happens long after the big bang and can at best give indirect indications for the form of the universe before it. There are other harbingers not of an electromagnetic nature, with a potential to show us traces of the big bang. With suitable information carriers interacting with matter less emphatically than electromagnetic radiation, one could possibly see back to earlier times. On the other hand, these messengers must be sufficiently sturdy and long-lived so as not to decay on the long trip to us. Only two possibilities then remain: neutrinos and gravitational waves.

Neutrinos are elementary particles that, in contrast to electrons, are electrically neutral and have almost no mass. They are easily created in radioactive decays and also played a role in the early universe. Unlike photons, the carriers of electromagnetic radiation, neutrinos are hardly absorbed by matter. One can see this impressively by the fact that neutrinos produced by nuclear fusion deep in the sun can reach Earth, while light can reach us only from the surface, where the temperature is too low for fusion. Most neutrinos even fly right through the whole planet: One can see them on the night side as well as the day side facing the sun, in clear contrast to light. On the other hand, owing to this extremely weak interaction with matter we notice only a small fraction of all neutrinos. From known production rates of neutrinos we can estimate that 1 cubic meter of space, everywhere in the universe, including on Earth, contains about 30 million of those particles, of which we notice nothing. Due to their small mass, all neutrinos produced in the early universe constitute, despite their vast numbers, only about 1 percent of the total density of the universe.

Thanks to this weak reaction rate, neutrinos can reach us from much earlier times than light. While light could freely penetrate space only thousands of years after the big bang, neutrinos were allowed to do so after about just a second! With a neutrino telescope, one could thus watch much earlier times in the universe, but the weak interaction of neutrinos with matter is a double-edged blade: for this very reason, neutrinos are difficult to detect on Earth. A neutrino telescope allowing us to see a significant fraction of all neutrinos, and even to determine the direction of their origin, remains a fantasy with existing technologies.

As a second alternative, we have gravitational waves: small perturbations of space-time that propagate with the same speed as light. They are generated, for instance, in collisions of heavy masses such as neutron stars or black holes; masses influence space-time and its curvature. Some of the curvature can split off from the collision range and advance into outer space like light leaving a star. The passage of a gravitational wave would announce itself by tiny periodic changes in distances between objects, in accordance with gravitational waves being propagating disturbances of space-time. If one could measure distances very precisely, one would be able to detect gravitational waves. Such detectors are currently being constructed in several places: the LIGO observatories in Louisiana and Washington, Virgo in Italy, GEO600 in Germany, and TAMA in Japan.

Very large masses are necessary to generate sufficiently strong gravitational waves. A direct detection of typically expected signals on Earth would require a precision length measurement of just a thousandth of the proton radius. As impossible as this may sound, clever constructions are already within reach of this goal, and they are being refined steadily. A direct detection is indeed expected within the next few years.[10] While this has not yet been achieved, there is little doubt as to the existence of gravitational waves. As described in the chapter on general relativity, the energy loss of double pulsars in close orbit can be explained precisely by an emission of gravitational waves as it follows from Einstein's equations. This is one of the closest agreements between theory and observation in all of physics. But the way to a gravitational wave telescope will be long and arduous. There are plans for a satellite system, LISA, which could fulfill such a purpose, and also for an underground telescope called ET—the Einstein Telescope. Realizing these plans will have to wait several years, and for a direct view deep into the big bang, even these systems will not be sensitive enough. Still, the prospect of a new kind of astronomy, entirely independent of the conventional one based on electromagnetic radiation such as light or radio waves, is an impressive one.

In the cosmology of the early universe, the big bang's high energies pose strict limits even to the propagation of neutrinos and gravitational waves. At such high energy densities as prevailed at those times, the universe is opaque even for those weakly interacting messengers. We are still denied a direct glance at the universe before the big bang, but an indirect investigation is not precluded. An earlier collapse phase, in combination with the repulsive forces of quantum gravity active during the bounce phase, should have weak implications for the ensuing expansion. Such indirect evidence may leave traces in the late phases of the big bang, which can be computed with theoretical models and possibly confirmed by comparison with observations.

Sensitive indirect effects require a detailed understanding of the theory and precise solutions of its equations. At present, even computer programs are not advanced enough to provide reliable values for the expected effects. But with further progress of the theory, combined with ever more precise measurements, indirect observations will come close to being possible. We will follow up on this further in the contexts of observational cosmology and cosmogony.

COSMIC FORGETFULNESS: FOGGY MEMORIES

What has he with all his grander might
done to lessen our country's plight?
A kingdom of soldiers he wanted to found,
to kindle and conflagrate all the world's ground,
to hold everything in his tight-fisted bound.

—FRIEDRICH SCHILLER, *Wallenstein*

What do theoretical considerations tell us? Independently of observations, theories are sometimes so strict that some possibilities, conceivable otherwise, can be ruled out. Although the equations of quantum gravity remain incomplete—those theories, after all, have not yet been fully formulated—possibilities can be constrained by mathematical consistency conditions. If we knew the present state of the universe precisely, we could calculate back to a time before the big bang, using all mathematically possible equations for quantum cosmology. From different equations one obtains different results, but if they do not differ too much in the whole set of possible equations, one could obtain a good impression of the state of the universe before the big bang. In a way, this resembles what is done by historians, as laid out for instance by Friedrich Schiller in his inaugural lecture[11] at the University of Jena:

> From the whole of these happenings the universal historian picks out those which have had an essential, unequivocal, and easily followed influence on the current form of the world and the state of generations now alive. The relation of historical data to the current constitution of the world thus is what must be sighted to collect material for world history. World history thus starts from a principle strictly opposite to the beginning of the world. The actual sequence of events descends from the origin of things to their most recent order; the universal historian advances from the newest state of affairs upward toward the origin of everything. When he ascends in thoughts from the current year and century to the preceding ones, and among the events offered by the latter, which contain insight for the next ones—when he has continued this walk step by step to the beginning—not of the world, for thereto leads no landmark—up to the first of statues, then it is up to him to turn back on the ventured way, and to descend unhindered and light along the guideline of the described facts from the first statues to the current age.

The parallel to history should not come as a surprise, for cosmology does investigate a kind of history: that of the universe. Here, one can push backward to much earlier times, perusing entirely new kinds of statues: cosmic background radiation, the distribution of galaxies in the universe, and supernovae (which will be discussed in the next chapter). In cosmology, this approach is known as "top-down": One starts with knowledge at the top end of the time axis and tries to derive properties at an earlier time at the bottom. Compared to that of Schiller, the view is upside down, because traditionally time in mathematical diagrams is drawn upward (or sometimes to the right). In cosmology, this approach may first have been introduced by Jim Hartle and Stephen Hawking, and was recently formalized by Hawking and Thomas Hertog.[12]

We do not know the present state very precisely due to the difficulty of observations of the entire universe, and we can never know it arbitrarily precisely due to quantum theoretical uncertainties of the cosmological wave function. The decisive question thus is how much of an influence the ignorance of the present state of the universe has on that before the big bang, obtained by calculating backward. Or: How many statues can we find in the early universe, in addition to the statuesque cosmic background radiation? Ignorance could be roughly constant in time, allowing acceptable precisions in our knowledge of the pre–big bang behavior; compared to the present state, ignorance would be increased only by the imprecisely known quantum cosmological equations. But it could as well happen that even a fixed form of the equations would show a significant growth in uncertainty toward the past. Such a behavior is known, for instance, from systems justly called chaotic: Starting from very similar initial values, their temporal evolution leads to widely differing results after only brief times. Examples are turbulent systems in hydrodynamics—the theory of liquid flow—in which chaos is the reason for the unreliability of long-term weather forecasts.

Chaos does play a role also in cosmology, but we can ignore it for the main questions of interest here. In the simplest models of general relativity, the total volume changes in a way that does not show chaos. Hope thus remains to venture backward successfully, at least in thoughts, but we still have to see whether the equations of quantum cosmology themselves might not add more uncertainty compared to those of general relativity.

Of particular interest is the question of what kind of quantum uncertainty could have prevailed before the big bang. Repulsive forces of quantum gravity prevent the singularity, save space-time from its demise, and make possible a world before the big bang. But as a quantum theory, quantum gravity also shows such typical properties as fluctuations and uncertainty. To some degree, the behavior can remain prevalent even at a distance from the big bang. One normally envisions a nonsingular space-time at the big bang as a so-called "bounce" where the universe reaches its minimal volume. But in reality one is talking about a quantum version of space-time, for without quantum theory there would be the singularity.[13] This must lead to fluctuations, alien to the classical world. Quantum forces are welcome, but they bring along unclassical jumpiness. Does the quantum world then keep space-time in fluctuating turmoil after having prevented the singularity—like an army of mercenaries that, long after repelling a dangerous aggressor, marauds through and plunders the ravaged lands?

In a precise analysis of currently available models one sees that many properties of the universe can be computed backward quite precisely. Many, but not all. There are physical quantities, important for some of the interesting questions, whose values before the big bang have such a weak influence on what is happening afterward that they are irrelevant for the present state. In such a case, the values before the big bang cannot be restricted by conditions one would obtain by extrapolating backward from the present values.

The universe, as it were, forgets which precise value such a property had taken before the big bang. Even though the value might in

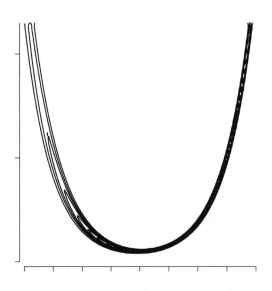

17. Illustration of a wave function for a nonsingular universe. Contour lines show the values of the wave function depending on the volume (vertical) and time (horizontal). The bottom axis, corresponding to vanishing volume, is never reached, and neither is the singularity. Quantum fluctuations, represented by the width of the ridge, may have been different before minimal volume was reached than they are afterward. This is the basis of cosmic forgetfulness, for the strength of quantum fluctuations before the big bang cannot easily be reconstructed after the big bang.

principle still have an effect on the behavior of the universe, the influence is so imperceptibly small that the value is irrelevant. No other phenomenon would still register it—it is forgotten. The main example of such forgetful properties is the degree of quantum uncertainty. Thanks to repulsive forces of quantum gravity, the big bang is no longer singular as in general relativity, yet it remains a rather special place. The value characterizing quantum fluctuations for a long time before the big bang suddenly becomes relatively unimportant after the big bang; quantum fluctuations are then determined by a new parameter, in general independent of the old one. Although the old value still influences what happens in a very minute way, by realistic observations it can no longer be recovered.

In other words, the state of the universe could have differed considerably in these undetermined properties from what we currently see. Their physical meaning reveals the quantum theoretical nature of the universe: While classical properties such as the volume or the expansion rate could be determined quite precisely, this is not the case for the quantum mechanical uncertainty of these quantities. In principle, the universe before the big bang could have been much less distinct than it is today, or its volume could have fluctuated much more. Thus we cannot be sure about the exact state whence the universe as we know it came. Within science, this does not appear more constrainable—unless one would allow oneself to make additional, prejudiced assumptions about the state. Although quantum cosmology can clarify questions about the universe in unprecedented ways, it does leave modest freedom for the myths.

LIMITS: NO ULTIMATE ANSWER?

The limits of my language are the limits of my world.

—LUDWIG WITTGENSTEIN, *Tractatus Logico-Philosophicus*

Loop quantum cosmology has shown mechanisms to prevent the collapse to singularities by counteracting forces. Results available so far are based on models simple enough to be treated precisely by mathematical methods; otherwise the described behavior would not have been revealed. Such a procedure is common in theoretical physics, in which one often starts with the simplest model characterizing a situation, only to make it more realistic by systematic extensions. In quantum cosmology, one might, for instance, consider different matter types or use fewer symmetry assumptions. In such a situation, the unavoidable question arises whether results obtained in the simple model are generally valid, or need to be revised in extended models. This general issue must also be taken into account for statements in quantum cosmology, such as those about repulsive forces or the associated cosmic forgetfulness, both seen in some models but not yet proven in general.

In this context, there is a considerable difference between negative statements about what can be known, such as cosmic forgetfulness, and positive results such as are often found in model systems. An example of the latter would be the specific density in a simple model universe at the time when it bounces. Initially one is often interested in such concrete data. In simple situations many computations can be done, but the question always remains of how reliable the result is under realistic conditions, which are more difficult to incorporate in the model. With cosmic forgetfulness, by contrast, we are dealing with a negative property: a result about limitations of knowledge. Such results may be less concrete, but they are more secure against extensions of models. Any extended version would, after all, only complicate the original model; otherwise one would directly have taken the more complex model as the starting point of one's analysis. And with a more complex situation, knowledge can be gained only in more difficult ways than in the simpler model. If knowledge was already limited, it surely will be even more so in the more general situation. It is thus safe to consider cosmic forgetfulness as a general property.

For physicists, limits to knowledge are depressing. The success of the deterministic worldview only strengthened the common belief in it; challenges are viewed skeptically. Cosmic forgetfulness is such a challenge, and it has been criticized. But most counterarguments so far deployed constitute acts of Zenoic desperation, abusing infinity, and are therefore invalid. One argument goes as follows: It may well be that the total uncertainty of the universe before the big bang cannot be determined, but there are many more, even infinitely many, properties characterizing a quantum universe. One uncertainty parameter out of infinitely many properties is almost nothing, and so nothing is being forgotten.

That this argument is nonsensical is obvious when it is applied to human forgetfulness: It is like saying that we all have perfect memory! It may well be that once in a while we forget something. But there are, after all, infinitely many possible pieces of information that we do not care to memorize and that, strictly speaking, we do not forget. And finitely many out of infinitely many ones is almost nothing; we remember everything. Also here the abuse of infinity is palpable, but in much more clumsy a form than in Zeno's case. Moreover, the argument is not valid for another reason: There are, in fact, infinitely many properties of the state of the universe before the big bang that are not reconstructible today. The degree of uncertainty is determined by just three parameters, one of which is relevant today, another one in the big bang phase, and the third before the big bang. Realistically, we can only hope to determine the value of importance today, and so the universe forgets two out of three parameters for fluctuations.

Although a quantum theory of gravity can avoid the big bang singularity—already a success—this does not necessarily mean that one can now determine everything in the universe and its exact origin. Doing so would in fact be far from possible: the universe still presents its riddles. In theoretical cosmology one often considers simple model universes in which the history and parameters such as the energy density or temperature at the time of its minimal extension can be calculated precisely. If special properties arising from the simplification are responsible for this high precision, results cannot be considered valid in general.

Even if a whole class of models provides similar statements, one cannot be certain about their generality. One has to be aware of selection effects, as described intuitively in Arthur Eddington's fishermen parable: Two fishermen chat about their job while their nets are filling with fish. One of them mentions a remarkable observation he has been contemplating: While the fish they catch have different sizes, no fish smaller than about half an inch has ever been seen in the net. The second fisherman does not find this enigmatic at all, for a fish grows in an egg before swimming freely in the sea and possibly being caught in a fishing net. He concludes that all kinds of fish must hatch with a size of about half an inch.

This is an example of an apparently valid but wrong conclusion. Obviously, the fishermen have not taken into account the fixed mesh size of the net, which can contain and pull on board only fish above a certain size. Smaller fish simply fall through the mesh or swim out of the net. Here we are dealing with a selection effect, for the method determines a property of the result. Similarly, there are selection effects in cosmological models whose difficulty in general requires specific properties for a detailed analysis in special cases. It is then difficult to decide whether the requirement of direct tractability itself, for instance by mathematical solutions, already determines at least some part of the result. One can analyze different models, but a large number of characteristic cases is necessary to arrive at a reliable result. In only a few instances is this possible, given the complexity of even the simplest models of quantum cosmology.

Instead, there is a rather more powerful view, which, however, is less popular. Abandoning the focus on positive properties, such as the precise value of a certain quantity, one takes a pessimistic stand, trying to show what limits exist for determining a parameter. In simple models one can compute these limits in much the same way as positive values, but results for limits are often wide and not very constrained. In some cases, however, such as the quantum fluctuations of the universe before the big bang, there are surprising limitations that reveal themselves even in simple models. Such limitations only become stronger in more general situations, and thus can be considered as general properties.

One should not misunderstand this pessimistic viewpoint as giving up. Limits play important roles in science, and they must clearly be recognized. Only then will one see how one can achieve progress within these limits, or whether it might yet be possible to evade them.

Positive values can be computed in simple models, sometimes easily, sometimes with considerably more effort. There are more models than one can imagine in which some computable parameters can be found, much appreciated by researchers as rich sources of scientific publications. Especially in the beginning stages of a newly emerging field, many possibilities for concrete calculations remain open. But the flood of data does not necessarily foster a deeper understanding. Often, known results are simply reproduced in a slightly changed model, but then one still remains ignorant about whether such a property is generally valid or the result of a selection effect. A transition to the pessimistic view, as recently initiated for loop quantum cosmology, is thus a sign of the maturity of a scientific field. One starts to take limits seriously, and turns to more general results within the studied theory. Besides, addressing and admitting limits is an important contribution to the honesty of science.

5. OBSERVATIONAL COSMOLOGY
PROBING THE WHOLE UNIVERSE,
PAST AND FUTURE

Any theory about space and time must be consistent with what we can see of the whole universe. Modern cosmology started in the 1920s with pioneering work by Edwin Hubble, but it remained an imprecise science for many decades. For most quantities characterizing the form of the universe, such as its expansion rate (now called the Hubble parameter), estimates were made. But according to the available data, they could easily have been changed by a factor of two or so. There were times when certain stars were estimated to be older than the age of the whole universe. With refined methods and advanced technologies, the situation eventually changed. Especially in recent times, observational cosmology has managed to achieve impressive progress, providing valuable contributions to an understanding of the universe. Today, it is one of the most exciting parts of physics.

Cosmology plays a special role within physics because the universe represents a unique system. It has not been prepared by an experimentalist, and so the usual procedure—generating many similar systems, such as groups of elementary particles in numerous accelerator reactions, to measure their properties repeatedly—is precluded. The common methodology of experimental physics allows one to considerably reduce the influence of mistaken measurements, which are bound to happen in a single unrepeated and unchecked experiment. With the universe, the situation is certainly different: Cosmologists must accept what the universe once and for all presents them.

There is only one branch in cosmological experiments: that of observation. Preparation cannot happen. Some observations are thus fraught with errors that cannot be eliminated completely. It is not always clear whether a measured value is the way it has been seen just by chance, or for a good reason. With repeated experiments one can easily detect and eliminate the role of randomness because it cannot give rise to systematic effects. Repetitions, however, are impossible in cosmology.

Cosmological parameters often have such surprising values that they almost cry out for a deeper explanation. For instance, as we will soon see, the universe has accelerated its expansion relatively recently, as inferred from the escape velocities of distant star explosions. It is sometimes speculated, perhaps in jest, that there may be a connection between the acceleration and the emergence of intelligent life able to perceive it. But how can one rule out that this is not just a result of chance? Making a decision about the role of chance becomes more complicated by the fact that the reason for the acceleration (mysteriously called "dark energy") is barely understood.

One should keep in mind, however, that many such coincidences exist, to which at times enormous importance has been attributed. For instance, the lunar and solar radii correspond so well with the radii of the earth and moon orbits that nearly perfect eclipses become possible. With other ratios, the moon could be too small to cover the sun completely, being hopelessly overwhelmed by the sun's brightness; or it could be too large and would demote the sun to a mere background player during an eclipse.

In fact, the apparent sizes of the moon and the sun in the sky are nearly equal, and impressive eclipses can result with a visible solar corona. As is well known, such events have often played important roles in ancient cultures. And by stimulating the desire to predict future eclipses, they have fostered support for astronomical research. Today, however, there is no doubt that the correspondence occurred by chance. Moreover, the moon is slowly moving away from the earth because the tides caused by its presence in orbit transfer energy from the earth to the moon. (Cosmic expansion does not play a role in this context, since its effects are overwhelmed by the gravitational binding of the earth-moon system.) The agreement of the apparent sizes of the moon and the sun is thus doubly random, the result of the unimportant properties of the earth and moon orbits in the solar system as well as of the period in time when we do our observations.

Issues are similar with many examples in cosmology. But starting a few years ago, observational cosmologists have become very clever in perusing various kinds of complementary measurements, and by now achieved unprecedented precision. This is the topic of the present chapter, with an emphasis on properties related to quantum gravity.

A further difference between measurements in cosmology and in the rest of physics is the fact that a cosmologist as observer is always part of the system under investigation: the universe. In other branches of physics, by contrast, the observer is separate from the prepared and measured system (except for rare self-experiments). Especially in combination with quantum physics, or quantum cosmology, this often leads to difficulties in understanding the theory, and to apparent paradoxa. We will come back to this issue in chapter 9, which addresses the uniqueness of cosmological solutions.

THE TRIAD OF OBSERVATIONAL COSMOLOGY:
A MIGHTY TRIUMVIRATE

Cosmological insights into the large-scale structure of the universe and its expansion are currently founded on three rather different kinds of observation:

(i) measurements of the already mentioned cosmic background radiation;

(ii) wide-ranging mappings of matter as it is collected on large scales in galaxies;

(iii) precise determinations of distances and velocities of violent star explosions called "supernovae type Ia."

In addition, there are data on material decomposition, especially the relative abundances of light elements such as hydrogen and helium, that give information about an early phase in the universe called *nucleosynthesis.*

THE COSMIC MICROWAVE BACKGROUND: HEAVENLY DISTRIBUTIONS

A painting of you: not on the wall,
on heaven itself, over it all.

—RAINER MARIA RILKE, *The Book of Hours*

Electromagnetic radiation once emitted as light has been traveling through space since early times, starting less than half a million years after the big bang,[1] quite early in comparison with the total age of the universe of slightly less than 14 billion years. The radiation is rather faint, but it can now be measured in many details such as the distribution of its varying intensity over the sky. As already mentioned, this radiation had its origin in the hot universe near the big bang and began its travels to us when matter became transparent, at a temperature of about 4,000 degrees Celsius. At that time, matter had not yet been able to concentrate in galaxies despite gravitational attraction, for the radiation pressure of the hot universe would immediately have pulverized any denser aggregate.

Although they were emitted very early in the universe, the electromagnetic waves still exist as so-called background radiation. Unlike directed radiation from a single burst, such as a star explosion, which would eventually pass us and quickly fade away, the background radiation lingers like noise in a crowded room. Similar to heat in a hot air balloon, it pervades space, simply being diluted, and thus cooled, when the universe expands. Its wavelength is stretched out by the expanding space—it is no longer visible as light but still detectable as microwaves—without ever disappearing. The background remains with us, and presents revealing views of the universe.

In the very early universe, the distribution of matter as well as radiation was almost homogeneous: nearly equal at all places. Slight fluctuations existed nonetheless, seeds that at later times, after the radiation pressure decreased, led to the buildup of much denser regions, eventually culminating in galaxies. Traces of these small inhomogeneities are still discernible from the directional distribution of cosmic background radiation: Its intensity varies slightly from different parts of the sky. Such variations can be computed by means of theoretical models for the evolution of the universe at the big bang and then compared with measurement data to give information about the validity of the theory.

As early as 1948, the existence of cosmic background radiation was predicted by Ralph Alpher and Robert Herman, who estimated the temperature, surprisingly precisely, to be about 5 Kelvin, or −268 degrees Celsius. At that low temperature, the radiation was considered undetectable by most physicists, and so the prediction had little influence. It took twenty more years for cosmic microwave radiation to be detected, first only by chance in 1965 by Arno Penzias and Robert Wilson, as a result of new developments in cooled detectors.[2] Penzias and Wilson were awarded the Nobel Prize in Physics in 1978 (together with Pyotr Kapitsa for his work in low-temperature physics).

Owing to the limited measurement precision available at that time, the background radiation appeared to be very homogeneously distributed over the sky. Contributions to the intensity distribution from different microwave frequencies were, as expected, in close agreement with Planck's formula for black body radiation, already encountered in quantum mechanics. In fact, the universe itself can be considered a perfectly enclosed box: There is clearly nothing outside it. Nowadays, the microwave background is the most precisely measured example of black body radiation. Since the distribution depends on the temperature of the radiation, one can use it to measure the average temperature of the universe at those times. For the microwave background, this yields 2.7 Kelvin, about −270 degrees Celsius, very close to the prediction by Alpher and Herman. This sounds cold, even though the radiation is supposed to have come from the hot early universe. But it has only cooled down during subsequent expansion of the universe and was much hotter at the time of its release: about 4,000 degrees Celsius.

In 1992, the satellite COBE (Cosmic Background Explorer, launched in 1989) took another look at the microwave sky and provided the first measurements of the cosmic background radiation to show irregularities in its directional distribution, or anisotropies. Variations are very small, about a millionth of the total intensity, but measurable nonetheless. For this finding, together with the confirmation of Planck's formula for the background radiation, the principal investigators of the measurement apparati, John Mather and George Smoot, were awarded the Nobel Prize in Physics in 2006. In the meantime, numerous further experiments have been undertaken, initially on balloons in Antarctica. Today, satellites such as WMAP (the Wilkinson Microwave Anisotropy Probe) are providing spectacular data. A new satellite, called Planck and prepared by the European Space Agency (ESA), was launched in May 2009. It should drive the data to unprecedented levels of detail, and for several years will probably be the ultimate measure of all cosmological observations. (Given the time required to collect and evaluate data, first results are not expected before 2012.)

Anisotropies in the heavenly distribution of microwave radiation contain a gargantuan wealth of information about the physics of the early universe. They show how gravitational attraction led slightly denser than average regions to grow, some time between the hottest phase of the big bang and the moment when radiation was released in a sufficiently cooled-down universe. Compared to the total time passed since then, the allotted time until the release was not much; there was not much clumping, and so the density variations and anisotropies in the leftover radiation are small. But they are not zero, allowing crucial tests of our theoretical understanding. Using Einstein's equations, which tell us how matter behaves in an expanding universe, one can start with the measured values to determine matter distribution at even earlier times. The result can then be compared with theoretical models, allowing predictions for how matter might have been distributed during the very hot phase of the big bang—close to the realm of quantum gravity.

All this involves extrapolations and is indirect, preventing the crystallization of a unique theoretical model. Still, the data are already precise enough to rule out some ideas and support others. In closest agreement with observations is a large class of models summarized under the name "inflation." They posit that the universe expanded extremely rapidly for some period of time in the very hot phase, even accelerating the rate of its expansion. (As already mentioned, the universe entered another accelerated phase relatively recently, which is independent of the inflationary acceleration.) Accelerated expansion of space, an apparent repulsion of all masses in the universe, seems in conflict with the attractive nature of gravity, and so the physical mechanism causing inflation remains hotly debated. We will later come back to this question in more depth.

Further details of the anisotropies give insights into some parameters of importance for a characterization of the cosmos. For instance, the anisotropy spectrum—obtained by viewing the intensity distribution on the sky as a sum of single uniform oscillations—has one clearly dominant contribution. Its wavelength is a measure of the curvature of space given at a fixed time. Space-time and its spatial part do not have to be flat and planar, but could be curved like a sphere. In this case, space is said to have positive curvature. A planar space, by contrast, is flat and has zero curvature, while there is also the possibility of negative curvature whose space is saddlelike.

Curvature in general relativity now causes what we perceive as the gravitational force, also acting on light or the microwave radiation in the cosmic background by bending their rays. Depending on the curvature, radiation traveling to us since its release is influenced differently by the space it traverses; it may be focused or unfocused. When we watch the sky in its microwave background light, positive curvature acts like a lens held in between, magnifying the intensity variations and shifting the wavelength of the dominant intensity variation. From the measured wavelength, one finds that the spatial curvature must vanish almost completely: The space we live in is nearly flat on large scales. (To be sure, this refers only to the curvature of space, as a part of space-time. Space-time as a whole is curved by the matter it contains; only an empty space-time would be completely flat. Moreover, the current universe is spatially curved on smaller scales owing to the presence of large masses and their strong gravitational force; only on large spatial scales is space flat on average.)

From smaller hills beside the main maximum of the anisotropy spectrum—hills whose presence was already predicted in 1965 by the physicist and human rights activist Andrei Sakharov—additional properties can be inferred. Of interest is the total amount of matter that led to the gravitational clumping. On its own, this parameter does not mean very much, but it is important in comparison with the amount of matter needed, according to general relativity, for a universe with an almost flat space. Curiously, the value of the amount of matter is much lower than expected for a flat space, even though one can infer from the position of the main maximum that the space is flat.

Data gained from the hills of the anisotropy spectrum thus seem to contradict each other—unless there is another, as yet unknown form of energy that does not lead to clumping; it would be relevant

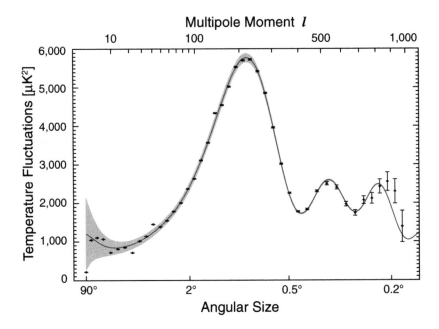

18. Spectrum of the cosmic microwave radiation with a precisely measured main maximum as well as smaller hills. (Image courtesy NASA/WMAP Science Team, http://map.gsfc.nasa .gov/media/080999/index.html.)

for the spatial curvature and thus the main maximum, but not for smaller hills. The disagreement of interpreted data looks like a severe problem that can be solved only by audaciously postulating a new form of energy, as yet unseen. Worst of all, observations require this energy form to make up about 70 percent of all the energy in the universe! Simply postulating any new form of energy, especially one so dominant, surely requires good reasons; otherwise it could hardly be taken seriously. Cosmology has by now provided a strong set of arguments, based on the fact that the new energy form is required not only by cosmic microwave data but also by other observations discussed in what follows. So the set of all observations is indeed consistent with the postulate of a new form of energy called dark energy, but this is a success paid for by the enigma of the energy's nature. So far, no satisfactory theoretical explanation has been found despite numerous attempts.

Adding insult to injury, the value of all mass not attributed to dark energy is decidedly larger than the amount of matter seen in the form of stars. Following the familiar script, this leads to another postulated energy form: dark matter, as distinguished from dark energy. It is rather prevalent, too, accounting for about 26 percent of the energy in the universe, but unlike dark energy, it does not cause accelerated expansion. For the matter we know, and are made from, only a meager 4 percent remains, a tiny minority. Indications for dark matter existed much longer, inferred from the velocities of stars in galaxies. So it is heartening to see that very different kinds of cosmological observations lead to consistent results, despite the darkness. Dark matter will not be discussed here much because it has few connections to quantum gravity. As dark matter candidates, objects as diverse as black holes and new kinds of elementary particles were proposed; the latter, if they exist, may possibly be found in accelerator experiments.

Another interesting characteristic of background radiation is its polarization. Electromagnetic radiation is of transversal form: Like a rope, it can swing in different spatial directions perpendicular to its own axis. Also the plane in which this swinging occurs, together with its variation over the whole sky, can be measured for cosmic microwave radiation. Recent new measurements by the WMAP satellite have shown that the polarization exists and have indicated the first concrete, though not yet precise, values. New data from the Planck satellite should provide an additional wealth of information about background radiation. Polarization can be caused only by the periodic space-time deformations of gravitational waves, stretching the stage for microwaves to move on, but not by other kinds of space-time curvature as they come from matter. From polarization data one can thus draw conclusions about the intensity of gravitational waves near the big bang, and further test or even constrain theoretical ideas such as traces of quantum gravity.

GALAXY MAPS: MILKY WAYS BY THE MILLIONS

> There one begins, a tiny wretch
> So happy smallest bites to catch,
> Thus one is growing patch by patch
> And practices to play a grander match.

> **GOETHE,** *Faust*

Sometime after the release of the cosmic microwave background, galaxies formed out of those very density variations that are still recognizable by their seeds in the slightly nonuniform background radiation. Initial irregularities in matter distribution were small, but denser regions did exist. By exerting a stronger gravitational pull on the surrounding matter, they attracted still more matter. Sometimes adjacent density centers came close and merged, the most massive and greedy ones swallowing almost everything in their neighborhood. Over time, the centers swelled more and more. They gained mass and became denser, and their internal pressure and temperature rose; eventually, stars ignited in them. These accumulations culminated in impressive galaxies, while the impoverished intergalactic region was deprived of matter.

Since Einstein's equations allow one to compute the clumping of matter, starting from an initial configuration as can be inferred from the background radiation, a precise mapping of galaxies provides a further test of the theory. Such wide-ranging mappings have become possible with ever more sensitive telescopes detecting even distant galaxies and their positions. Although existing maps remain incomplete, the resulting images, in particular those of the Sloan Digital Sky Survey (SDSS), are impressive (see figure 19). The SDSS includes millions of galaxies in our cosmic neighborhood, showing indeed how homogeneously matter is distributed in the universe. By comparison, what our naked eyes show us in the sky around Earth is only meaningless detail—a historical chance event in the development of the universe and of our own existence.

This large-scale homogeneity turns out to be an immense stroke of luck for cosmology, for such a high degree of symmetry implies strong simplifications in solving Einstein's equations. Otherwise one could not see the cosmic forest for all the galactic trees in their complicated evolution, all observations notwithstanding. To model the universe, so many details would have to be taken into account that no satisfactory explanations could be gained. Fortunately, the universe does help (or possibly fool?[3]) us.

Besides this confirmation of homogeneity, long confidently assumed by cosmologists, galaxy maps provide concrete data supporting conclusions drawn from the cosmic microwave background. Despite the strong homogeneity over large distances, the distribution of galaxies shows small variations on different scales. As in the microwave background, one sees a clear maximum for a certain distance between galaxies: one is much more likely to find a pair of galaxies separated by that distance than by others, an observation

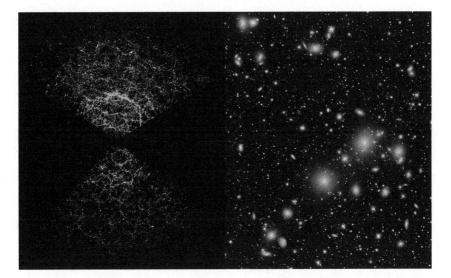

19. Galaxy map from SDSS. On large scales, the universe looks much more homogeneous than the close neighborhood of the solar system and the Milky Way would indicate. (http://www .sdss.org/news/releases/galaxies.jpg.)

announced simultaneously in 2005 by the collaboration of SDSS and 2dFGRS (the Two Degree Field Galaxy Redshift Survey). As before, Einstein's equations lead to a connection of that distance with the spatial curvature, again showing evidence for a flat space.

Such agreements of very different measurements, demonstrating the consistency of the data, always constitute impressive indications that we are on the right track toward understanding the universe. What is more, in this specific case it indicates that the spatial curvature remains constant on long time scales. After all, the tiny density fluctuations from the release time of the microwave background on the one hand, and the distribution seen in galaxy maps on the other, are separated by millions of years. Through all this time, the spatial curvature must have remained unchanged. That the spatial curvature is constant in time is a prediction of general relativity for a universe with nearly homogeneous mass distribution. Thus the theory has passed another test superbly.

SUPERNOVAE: THE WISE LAST WORDS OF DYING STARS

After the first galaxies formed, their stars evolved further. At some time, those living fast approached the end of their nuclear fuel; high enough inner temperatures and pressures to prevent gravitational collapse could no longer be sustained. For some of them, the collapse into a black hole was waiting.

This process and the structure of black holes will occupy us in the next chapter. Here we are interested merely in the fact that not all the starry matter participates in the collapse, a significant part being blown out into space. The released energy can be seen as a huge explosion, a *supernova*. One subclass of supernovae, called "type Ia," has the advantage of having very special properties: Precisely observing the color distribution of their emitted light, for instance by decomposing it with a prism, allows reliable conclusions about the total amount of energy spewed out. From this, in turn, one can derive the supernova's distance from us using two basic facts: The explosion should appear darker if it is farther away; and the more energy emitted, the brighter it should be. Its visible brightness thus depends on the distance and the total energy. Since we directly measure the brightness and determine the total energy from the color distribution, the distance can be inferred. For normal stars, without any information as to how brightly they are burning, this would be impossible. Their distances can often be found only indirectly, for instance by comparison with other objects in their neighborhood, such as those supernovae. For this reason, supernovae are designated standard candles, a kind of normalized lighthouse for the exploration of the universe by astronomers.

Moreover, supernovae shine very intensely as a result of the violence of the underlying event, making them visible from afar. Owing to the finite speed of light, they allow us to see traces of the universe a long time ago, at the time when their light had to start on its journey in order to reach us now. The expansion history of the universe, in turn, can be traced back by determining not only the distance of the supernovae but also—again from the colors of their light—their velocity caused by the expanding universe. (Additional contributions to the velocity are made by gravitational attraction to masses in their neighborhood. Since supernovae, like their progenitor stars, are situated in galaxies that tend to rotate around a center, this motion adds to the supernova velocity. But these contributions are usually small and, moreover, are distributed randomly compared to the systematic expansion velocity; they do not play a role when a large class of supernovae is considered.)

The distance thus provides the supernova's explosion time before now; the color distribution, the cosmic expansion rate back then. In this way, a relationship results: expansion rate depending on time, a function of exactly the form obtained by solving Einstein's equations for a given energy composition in the universe. Comparing a large class of solution curves with the observed piece of the expansion-time relationship allows one to see which composition best agrees with the recent past of our universe. This constitutes another method to determine cosmological parameters, independent of those measurements discussed in the preceding two sections. Here one has to see single stars, in contrast to whole galaxies in their mappings or even the sky-wide microwave radiation in its cosmic background. Supernovae, accordingly, allow only a determination of the most recent part of the universe's history. By referring to vastly different times, the main contributions in the whole of all obtained data—cosmic microwaves, galaxy maps, supernovae—are independent of one another; checking whether they agree in details is an immense consistency test.

Consistency is confirmed indeed, and it holds a surprise. We saw that details of the heavenly intensity distribution of microwaves can be explained only if there is a very special, dark contribution to the energy of the universe. In recent times, such a contribution must have implied an acceleration of the cosmic expansion. The recent past is just the domain of supernova observations, and the expansion history they provide shows the same acceleration. Such an energy form is thus seriously to be reckoned with, even though a theoretical explanation is still lacking.

Postulating new forms of matter to explain otherwise irreconcilable data is audacious, but nothing new in the history of physics. Wolfgang Pauli, for instance, in 1930 proposed to relate an apparent energy loss in certain radioactive decays to the existence of a particle unknown at that time. The particle was supposed to interact only weakly so as to elude direct measurements and be perceptible only by the energy it steals away. Everything was cooked up very nicely: The postulate of a new particle conveniently combined with a mechanism to make it unobservable can hardly be ruled out, or so it seemed. But not even Pauli liked this; although the energy loss was confirmed by more sensitive methods, he himself considered this explanation a last resort—an emergency treatment to save the law of energy conservation so important for all of physics. But Pauli was right: The particle was finally detected in 1956 by Frederick Reines and Clyde Cowan; Reines was awarded a share of the 1995 Nobel Prize in Physics for his part in the discovery (Cowan died in 1974). Despite the weakness of interactions, the detection was made possible by the strong source of a nuclear reactor;[4] given sufficiently many elusive particles, not all of them can escape unseen. We have already encountered this particle: it is the neutrino, whose name was introduced in 1934 by Enrico Fermi. It plays an important role in modern astrophysics and cosmology.

NEGATIVE PRESSURE:
ELUSIVE REPULSION

Will cosmology be granted a similar success: a direct and unquestionable proof of a new enigmatic energy form it suggested? Before this can become reality, theorists first have to hunt for possible explanations in the zoo of particle physics or the menagerie of gravitational phenomena. Without any such indication, an experimental search would resemble that for the proverbial needle in the haystack. Many exotic candidates are currently proposed as possible sources, all providing negative pressure. The nature of some of them will eventually bring us back to the topic of quantum gravity.

DARK ENERGY: THE UNKNOWN 70 PERCENT

Nature tends to hide herself.

—HERACLITUS "THE DARK," Fragment

Even imaginative theory has much to grapple with in dark energy, whose properties are far from common. After all, this energy form must cause accelerated expansion, and thus repel not only the remaining matter in the universe but even space and time. Such a behavior is highly unexpected for matter subject to gravity, which normally causes masses and energy distributions to attract each other—implying for instance the serious singularity problem from unchecked collapse.

Quantum gravity can lead to repulsive forces, but it is usually significant only on the tiny distance scales of the Planck length, at which energy can easily behave in unexpected ways. With dark energy, however, we see repulsion on the gigantic scales of the entire current universe. To explain this with quantum gravity is conceivable but extremely difficult. It can happen only if small quantum gravity effects in tiny spatial regions add up in just the right way for the observed behavior to occur. Concrete calculations, let alone possible direct detections of this new energy form, remain elusive. We must postpone quantum gravity a little longer.

Another possible solution is to doubt some of the cosmological premises. When evaluating data, one naturally assumes that a spatially homogeneous model with uniform density everywhere can be used. As shown by the galaxy maps, this is indeed a reasonable assumption. But the accelerated expansion began relatively recently, after the phase of the universe shown in the majority of the galaxy maps. Could it be that gravitational attraction, having worked ceaselessly in the meantime, has led to stronger matter clumping, such that the homogeneity assumption is no longer correct?

This proposal, developed particularly by Thomas Buchert, has the advantage that the validity of general relativity need not be doubted. Accelerated expansion would occur only apparently, as an artifact generated by squeezing an already strongly inhomogeneous universe into the mathematical corset of a homogeneous solution. On the other hand, the assumption of homogeneity that came along with strong computational simplifications in the theoretical modeling is dropped. Such doubts are, of course, always justified, for predictions merely relying on the availability of simple mathematical solutions are never valid. Simplicity appeals only to our limited and biased minds, not to nature. Solutions of simple enough form can be analyzed in all details, but this comes at the expense of abandoning some properties more difficult to model, yet possibly significant. Whether a model is reliable in all respects can be judged only in a more general setting within the theory used; in general relativity, the required equations easily become so complex that the behavior of their solutions is hardly known. So this proposal, alas, has not yet resulted in completely convincing explanations of dark energy.

It is perhaps consoling to know that accelerated expansion in certain phases is at least possible, though not natural, within general relativity, despite its generally attractive character. In Newtonian gravity, only the mass of matter calls gravitational forces into action, and those forces are purely attractive. In general relativity, by contrast, gravity is caused not only by masses but also by energy, which, according to special relativity, is equivalent to mass by Einstein's famous formula, as well as to its relative, pressure. Like most differences between Newtonian and Einstein gravity, the role of pressure is once again a consequence of the transformability of space and time. What is primarily relevant for gravity is energy density: some amount of energy located in a certain volume. Since special relativity shows that mass is equivalent to energy, the usual gravitational attraction of masses qualitatively results, as it does in Newtonian physics. But relativistically one can transform space into time intervals by changing one's state of motion. Just as space and time cannot be viewed independently of each other, but instead form a single object—space-time—energy density cannot be considered a single independent object either; length intervals are used in its definition. After changing the state of motion and partially transforming space in time, the apparently static energy density first takes the form of an energy flow: What to one observer looks like energy located in a region, another one in relative motion would see as energy moving in a certain direction.

The mathematical object describing the energy density is slightly different from that object giving directions in space-time. Space-time directions are called *vectors*, combining the spatial and temporal shifts. These are four components, corresponding to the four-dimensionality of space-time. Energy density and energy flow—the latter with three components because of its spatial direction—already make four quantities, enough to complete a vector. But the mathematical object describing energy is not a vector; it is a *tensor*—the so-called stress-energy tensor. A tensor of this kind in a four-dimensional space-time has not four but sixteen components—one might view it as a vector squared, or a matrix.

What are the remaining components of the stress-energy tensor? As the name suggests, these are elastic stresses in matter as well as pressure. Pressure is defined as the negative change of energy divided by the change of the volume enclosing the energy amount. This definition is in accordance with the intuitive behavior of matter under pressure: If one attempts to shrink the volume, one has to work against the pressure, thereby increasing the energy of the enclosed matter. The volume change is negative, the energy change positive; and pressure, defined as the negative ratio of these quantities, is positive. The more energy spent, the larger the pressure. But if the same amount of energy can achieve a larger change in volume, the pressure must be smaller. In this way the proportionality of pressure to the energy used and the inverse proportionality to the change in volume follow. Energy values as well as length changes also play a role here; it is thus understandable that pressure should be combined in one object together with energy density and energy flow.

In Newtonian physics, mass density causes gravitational forces. In a relativistic description one cannot separate energy density from the energy flow or from pressure and inner tensions. The gravitational force must thus be a consequence of the entire stress-energy tensor, as it indeed appears in Einstein's equations. Consequently, pressure is relevant for the expansion of the universe; the pressure, not just the density, of all matter contained in the universe contributes to its expansion history. Implications for the possibility of accelerated expansion are immediate: Like mass, the energy density must be positive and can play no role in the direct generation of repulsive forces. Pressure, by contrast, though often positive, can easily take negative values and then cause repulsive gravity—provided it is strong enough to overwhelm the positive energy density. Here we have found one of the main properties of dark energy: strongly negative pressure with positive energy.

Negative pressure is well known to arise in special situations. Negative pressure occurs, for instance, in the transport pipes of trees, caused by water evaporating at the tips of leaves. But what is its role in gravitation? As defined, pressure is characterized by the energy change for some volume change, with a conventional minus sign smuggled in to have a positive value in common situations. In this case, energy decreases when the volume is increased, and matter such as a gas will expand to lower its energy. To prevent this, the gas must be enclosed in a container with sufficiently strong walls. With negative pressure, matter can lower its energy by contracting, and only a pull can prevent this.

This appears to be the role of pressure as long as gravity is ignored.

For pressurized matter in a universe, one can again see the idiosyncratic trait of classical gravity, which already obliged us to cope with singularities in gravitational collapse. In that case, an immense piling up results from attractive gravity's compressing masses and further enhancing gravitational attraction. In general relativity, there is no escaping this unstable situation: After some time, self-destructive gravity ends in a singularity—unless it can be tamed early enough by repulsive forces of quantum gravity.

Gravitational pressure shows a similarly unstable behavior. Now we have, for instance in the homogeneous case of cosmology, matter "enclosed" not in a container but in the entire space of the universe. With positive pressure, matter would have to expand, which would be possible only by an expansion of the whole universe. But positive pressure, like positive energy, implies gravitational attraction, and the cosmic expansion is slowed down. The tendency to expand at positive pressure is counteracted by gravity. If matter under positive pressure is localized in a collapsing piece of space-time, like a burnt-out star, the collapse is accelerated, and at the end of this unstable situation waits the singularity of a black hole.

With negative pressure, the situation is just the opposite, provided it can compensate for positive energy density. Matter in the universe should then contract, which would be possible for a homogeneous matter distribution in the whole cosmos only with contraction of the universe. Gravity reacts adversely by enhancing cosmic expansion, and accelerated expansion results. As observations indicate, the expansion of the universe in the recent past was indeed accelerated. If the homogeneity assumption is justified, one has to expect an immense matter contribution of negative pressure.

Though it presents at least a theoretical characterization, this account is far from being an explanation. To impute sufficient negative pressure active over long cosmic durations one must deploy many tricks from the toolbox of theoretical physics. Negative pressure is possible; but not only must it be strong enough to compensate for the positive energy density of all matter in the universe, it also must prevail for extremely long, cosmologically relevant time scales. As of now, no phenomenon is known that would be able to manage this; current theoretical proposals remain unconvincing. Reducing cosmic acceleration to concrete new forms of matter appears, even more than in the case of Pauli's neutrino, as the last, reluctantly employed emergency solution. For dark energy, it is not sufficient to assume the existence of a new particle and perhaps postulate some value for its mass; many details of its precise properties would have to be contrived. So much apparently random input would be required that this kind of procedure could hardly be called an explanation. Here we have exactly the ominous problem of dark energy.

INFLATION: SPECULATING ON THE EARLY UNIVERSE

Negative pressure can suspend the purely attractive nature of gravity in general relativity, so could it perhaps be important to prevent singularities? For this purpose, unlike for dark energy, matter under negative pressure would have to be distributed not in a wide universe of current size but in the small early universe near the big bang. In the early universe we can be quite sure of the homogeneity assumption, thanks to the detailed observations of background microwaves. Moreover, densities and temperatures were much higher than they are now, possibly allowing quantum theory to play a larger role.

Negative pressure in the early universe is called *inflation*,[5] with an allusion to the inflationary — accelerated — expansion of the universe. While inflation's negative pressure could be suspected to help prevent the total collapse of a small universe seen in reverse, it is not strong enough. It does contribute a kind of repulsive force, but as shown (under certain conditions) a few years ago by Arvind Borde, Alan Guth, and Alex Vilenkin, it cannot prevent the singularity completely. Alan Guth was the first to introduce the theory of cosmic inflation in 1981, when he emphasized other advantages independent of the singularity problem. (At the same time there were similar, but not much noticed, articles by other authors.)

Inflationary research can be seen as a kind of template for quantum gravity: It remained speculative for decades, driven mainly by principles and perceived inconsistencies rather than by experimental input. Only in recent years has it come close to observational contact, and by now some models can already be ruled out. On a smaller scale — efforts were not as intense and prolonged as they have been for quantum gravity, and mathematically the field is much less challenging—we can thus observe how good a theoretical framework that is blind in its experimental eye can be. Inflation does have a major, mind-boggling consequence: It can explain how everything we know, including ourselves, arose from quantum fluctuations in an initially empty universe.

Although inflation has never been observed directly, it presents an elegant mechanism to explain other observations of matter distributions in the universe. It is also one of the most energetic phenomena known with at least some indirect access, and so it is of considerable interest as a tool with which to probe quantum gravity. Densities during inflation were not as high as the Planck density, but in typical scenarios they come close to at least about a millionth of its value. With the immense size of the Planck density—those trillion suns compressed to the size of a proton—this is as good as it gets. Consequently, models to combine the various approaches to quantum gravity with inflation, or even to derive inflation as a necessary consequence of some model, abound. Details and long-term consequences of such a quantum-gravity-induced early-universe episode could be used to test ideas for quantizing gravity.[6]

String theory, as always, has provided the largest variety of scenarios to possibly result in an explanation of inflation. This achievement is not at all trivial, for string theory, as a consequence of its symmetry principles, deals much more naturally with the opposite situation: negative energy density with positive pressure, rather than positive energy density with negative pressure. Once enough ingredients of string theory, related to what is going on in its extra dimensions, became understood, researchers were able to start constructing configurations able to account for the inflation of a universe. Finally completed by Renata Kallosh, Shamit Kachru, Andrei Linde, and Sandip Trivedi, such models were subsequently studied by many others. But in these or other scenarios, the constructions appear too contrived and not restrictive enough to provide clear predictions regarding inflation. In loop quantum cosmology, any inflationary effects would be a consequence of its basic properties, the atomic nature of space and time. Although the number of possibilities is smaller, it is still not small enough here to suggest clear test conditions for cosmological observations. Specific aspects will be discussed later in this chapter and in chapter 8.

So far, inflationary behavior of the early universe has been considered mainly in the context of more traditional questions in cosmology, in particular to explain how the large-scale structures we see now could have arisen. To see the role of the proposal of inflation, we first indicate what cosmology would be like if no inflationary episode had happened. Cosmic background radiation allows one to draw conclusions about early matter with its near but not exact homogeneity. Fine structures in this heavenly intensity distribution can be computationally evolved to earlier times by letting Einstein's equations run backward. In this backward evolution, the distribution becomes more and more homogeneous—which is not surprising, since it is understood as an indication of gravitational clumping building up in time. The earlier in the big bang phase we are, the more homogeneous the universe is. This looks attractive, for it indicates that our complex universe could have emerged from a very simple initial state. Only the reigning gravitational attraction of matter in an expanding universe would have led to all the complicated structures we can now see, in gigantic accumulations of galaxies all the way down to our solar system and the planets. A simple initial state suggests that we could one day, in an ultimate explanation, infer the reason for our own universe and answer the fundamental questions: Why is it as it is, and why is it at all?

Alas, a very homogeneous initial state in the early universe may present a severe problem for the consistency of cosmology. Viewing the universe with its big bang singularity as a starting point, there is a finite maximal duration for the propagation of signals between two points: If a signal starts at the singularity (or shortly thereafter, since all theory breaks down at the singularity itself), it can travel only a certain distance to some other point at a later time. Using the known speed of light, the largest distance can be computed in general relativity, and it is so small that normal circumstances would not allow light signals in the early universe to traverse all space. Here lies the problem—the horizon problem—for how could a homogeneous distribution, with equal densities everywhere, have come about if no signals could have controlled this homogeneity?

At this point inflation enters the game, because the mentioned "normal circumstances" include positive pressure as one condition. With negative pressure active for sufficiently long, on the other hand, it is, as Guth noted, very easy for signals to communicate between all possible places in the early universe. The rapid expansion of space-time would carry along all signals, allowing them to reach much farther distances not by their own motion but by surfing on the accelerating universe. No consistency problem exists in an inflationary universe, which is one of the attractions of the theory of inflation.

As one can see from the argument, the existence of the big bang singularity plays an important role, for it is taken as the starting point of the universe. But on account of the breakdown of the theory, the universe cannot be considered to have begun at the singularity. In a reasonable theory, the singularity is instead eliminated by mechanisms such as the repulsive forces of quantum gravity. Then, we automatically solve the horizon problem in a way different from that proposed by Guth: If the universe existed before the big bang, signals had more time, in fact as much as they needed, to connect all places in the early universe; a homogeneous distribution could easily be achieved.

As a motivation for inflation, the classical horizon problem thus disappears. Instead, inflation has a further, much more important advantage not yet seen by Guth's original considerations. Only the application of detailed calculations to determine how ripples in space-time propagate and evolve—based on methods worked out in 1981, mainly by Slava Mukhanov and G. V. Chibisov, even before the advent of inflation theory—showed that inflation leads to a special kind of very nearly homogeneous matter distribution. This form of the distribution is exactly what is also indicated by current observations. Although the accelerated expansion of the early universe, in contrast to that of the late one with dark energy, cannot be seen directly, there are strong, if indirect, indications for inflation, that is, for negative pressure in the early universe.

The explanatory power of inflationary models is much larger yet. They not only can show why matter is so homogeneous, but they also provide a mechanism for understanding why matter as we know it arose at all from the hot, dense, mangled big bang phase: In a universe expanding with acceleration, matter particles are continuously being created out of the vacuum—from perfect emptiness! As in our earlier description of the quantum mechanical wave function as ocean waves that are piled up strongly by a quaking ground, wave functions of matter are being excited by the "trembling" stage of an accelerated space-time. An initially empty vacuum state of matter is being filled more and more with real matter particles during inflation.

In cosmology, this means that even a state of perfect vacuum at the big bang does not remain empty as inflation goes on, but is increasingly being enriched with matter. For a sufficiently uniform expansion, one with almost constant acceleration, the matter distribution toward the end of inflation automatically takes a nearly homogeneous form, in excellent agreement with measurements of cosmic background radiation.

Such a process of matter creation is possible only if gravity is included in the equation; otherwise it would be in contradiction with energy conservation. The energy carried by newly emerging particles must come from some source; it cannot just arise from nothing. Negative pressure also plays a role here: Normally, when the gravitational force is weak, matter under negative pressure must become denser; it can lower its energy by contracting, and consequently it does so as much as possible. Only if matter is surrounded by sufficiently strong walls can this contraction be prevented. In the universe, matter is distributed homogeneously and uncontained; nothing should prevent it from contracting. But in its idiosyncratic way, gravity does not respect this tendency and instead reacts to negative pressure with accelerated expansion: With positive pressure, we expect matter to drift apart, but since pressure contributes to energy, it causes gravitational attraction. Negative pressure means a change of sign, reversing the effects: We do expect contraction, but instead, enhanced expansion occurs under gravity. Rather than being lowered, matter energy is increased, sponsored by gravity. All this becomes perceptible by the creation of matter excitations, or particles, even if the initial state was vacuous.

Explicitly, particle creation happens as follows (in this context, matter must be described by quantum theory, since only this allows a full understanding of its vacuum state). Like any state, the matter vacuum in an expanding universe has fluctuations: the number of all particles is zero on average—we are, after all, in a vacuum state—but not precisely so. This fluctuating behavior is analogous to the fact that the position of an electron's wave function is precisely determined only on average, while concretely it fluctuates within the spread of the wave function; the electron can be encountered at different places. Similarly, the matter wave function, in which the number of individual particles plays a role comparable to that of position in quantum mechanics, is spread out and does not have a precise particle number. If the universe did not expand, there would be no further consequences; the particle number would be imprecise but zero on average and remain so forever. But the expansion of the universe implies that, as in the case of effective forces in quantum mechanics, fluctuations influence the particle number as determined by the wave function of matter. In the course of time, new excitations arise, making the particle number nonzero even on average.

The concept of particle pairs is useful here. As is the particle number in a vacuum, energy is also imprecise, allowing the creation of pairs of matter and antimatter particles—such as an electron and a positron—for brief periods of time. Under normal circumstances, the partners in such a pair would immediately annihilate each other into radiation, but this can be prevented if the two particles are rapidly separated far enough. Handling individual particles directly and keeping them apart is certainly impossible, but when the particles are electrically charged, with the antiparticle always having the opposite charge from the particle, it could be achieved by subjecting the pair to a strong electric field. While such an experiment could in principle be performed in the laboratory, it would not explain the creation of particles in the early universe. At those times, no strong electric fields that could have achieved large-scale matter separations were active; otherwise the universe, and the distribution of its galaxies or background radiation, would be much more anisotropic as a result of the direction of the electric field.

Alternatively, one can separate the particles by, as it were, pulling apart the ground beneath them. This is exactly what happens in a universe expanding with acceleration: Space-time itself so rapidly pushes outward that particles and antiparticles emerging from the vacuum no longer annihilate each other. They simply have no chance of getting close, against the burst of space-time. Instead they remain as long-lived particles created from the vacuum. According to these inflationary ideas, the universe brought about matter, and consequently all structures now visible, in its initial nearly homogeneous distribution. In this picture, the entire matter content of the universe is a random event of quantum fluctuations, enhanced by gravity and frozen into real existing matter.[7] All that is necessary is a phase of accelerated expansion, or a matter form with negative pressure.

Here theory has to fight hard to provide a concrete form of such suppressed matter. Very special arrangements must again be made to achieve negative pressure for sufficiently long times. In the simplest and most elegant models of inflation, the negative pressure has to reign so long that the universe expands by a factor with twenty digits; otherwise no sufficiently homogeneous matter distribution in agreement with the cosmic background observations would result.

One can mathematically tailor such a matter form, but it requires very special interactions as well as initial values for that matter forced to be under negative pressure. This is similar to the production of negative pressure on Earth, which can be achieved, for instance, by supercooling an amount of matter. As matter usually expands when it gets hotter, it contracts when cooling down. When enclosed in a container, the contraction implies a pull perceived as negative pressure. While such conditions are possible, they are by no means common. On the one hand, special experimental installations are required; on the other hand, negative pressure is rarely sustained for long times. Heat exchange with the surroundings will quickly cause the supercooled matter to reheat, and the negative pressure disappears. How exactly, or if at all, the universe can produce negative pressure remains a riddle.

OBSERVING QUANTUM GRAVITY? DESPERATE FOR CLUES

In the universe one directly observes or infers not negative pressure, but accelerated expansion. Negative pressure is the only classical way by which general relativity can sometimes, in special situations, generate repulsive forces despite its normally attractive character. Primarily, however, we are interested in an explanation of the acceleration and of the repulsive force itself, and here, as in the case of the singularity problem, quantum gravity can come to our aid. If this works, intricacies encountered in attempts to justify negative pressure can possibly be avoided.

In contrast to dark energy, which must be explained for a big universe in which no significant role is expected for quantum gravity effects, quantum aspects of space-time may well have been important in the early and hot universe. In this phase, after all, a repulsive quantum gravity force, at least according to loop quantum cosmology, acts to stop the collapse of the preceding universe at the big bang and turn it into the expansion visible now. But a force that stops the collapse can also drive an already expanding universe to accelerated growth, just as a rock thrown up in the air first slows down and stops for an instant, then falls down back toward the ground with increasing speed. Effective forces of loop quantum cosmology, active when its discrete space-time repels rather than absorbs matter energy, indeed easily make a universe of small sizes expand in an accelerated way. No specially selected matter under negative pressure is necessary. Acceleration instead happens as a consequence of a new behavior of gravitation, a changed force that at high energies and small extensions of the universe deviates from general relativity. If the gravitational force is no longer purely attractive but has a repulsive component, even conventional matter can be in agreement with accelerated expansion.

So far, however, it remains unclear whether the degree of acceleration is sufficient. In any case, the horizon problem does not arise in the absence of a starting point such as the big bang singularity; but as in inflationary models based on negative pressure, one needs expansion by a sufficiently large factor to ensure a homogeneous matter distribution. How this is possible in general will be revealed by investigations currently in progress by several groups in the UK, for instance Ed Copeland in a fundamental study together with David Mulryne, Nelson Nunes, and Maryam Shaeri, as well as independently by Aurélien Barrau and Julien Grain in France, Shinji Tsujikawa in Japan, Gianluca Calcagni, and others.

These cosmological investigations rely on substantial derivations of equations to describe the evolution of small disturbances in the matter distribution. A latticelike structure of space-time leads to deviations from the continuum equations of general relativity, which can be evaluated and eventually compared with observations. Whether the involved calculations required could lead to success remained unclear for a long time. In general relativity, one is dealing with an overdetermined set of equations: There are more equations to be solved than free variables, a fact related to the main difficulty in finding a quantum theory of gravity. In general relativity, the set of equations is consistent, but if extra terms or effective forces are included as they may result from quantum gravity, this consistency is not easy to maintain. If two of the equations give different solutions for one and the same variable, no consistent common solution to all the equations can exist. Nevertheless, for one type of quantum correction, consistency has recently been realized in calculations performed mainly by Golam Hossain and Mikhail Kagan over three years. More than once in this process it seemed that all attempts would fail—a sad outcome that would have been especially devastating for Mikhail, who had based the main part of his Ph.D. work on these questions. Fortunately, all obstacles were successfully circumvented: A set of consistent equations is now available and under intense cosmological scrutiny.

Possible tests of the theory, in this case loop quantum gravity, then arise. The effects of quantum gravity are usually small, even near the big bang, but in some cases they can add up during long cosmological stretches of time. If this does happen, observations may indeed come within reach. Should that be realized, it would constitute a major milestone in modern theoretical research, for so far no experimental indications exist for the exact form of quantum gravity. Just as experimental data—for instance, atomic spectra—were always important guidelines during the development of quantum mechanics, observations will play an equally essential role in completing the currently discussed models of quantum gravity to a successful version.

In loop quantum gravity, observations, even if only indirect ones, would have another exhilarating advantage. They would give insights into the repulsive force that at the big bang gave rise not only to accelerated expansion but also to the turning point of the previously contracting universe to our expanding one. From precise observations of the repulsive force and its action during the big bang phase, we could draw conclusions about the ancestor universe itself. We would be granted a first, though indirect, glance at the universe before the big bang!

It is not only that the data of the cosmic microwave radiation have to be much improved, for which the Planck satellite would already be a large step forward, but also other independent sources of signals will have to be explored. Electromagnetic radiation simply cannot reach our measurement devices from sufficiently deep stages of the big bang phase; the more penetrating messengers of neutrinos or gravitational waves will be required.

In addition, these messages scatter, limiting how far back in time we will be able to see. With neutrinos, one can view a moment in time about a second after the big bang, much earlier in the universe than the slightly less than half a million years after the big bang that is granted by microwave radiation. But even this blink of a second does not penetrate deeply enough: A second still contains an unimaginably large number (extending to 42 digits) of discrete time steps of quantum gravity. Gravitational waves could come from even earlier times, but here quantum gravity still has to show clearly just how far back we can see. Even with gravitational waves, no arbitrarily distant hindsight will be possible. The principle of cosmic forgetfulness also limits a direct view of the preceding universe.

How, then, would it be possible to gain indirect insights into properties of the universe before the big bang? Here, the earlier comparison of the singularity with the splitting off of water droplets is useful again. The splitting off is similar to the big bang singularity and its prevention in loop quantum cosmology, with the continuous water drop playing the role of the smooth geometric rubber band of curved space-time in general relativity. The splitting off, in this view, presents a singularity, easily explained and resolved in an atomic theory of water—just as the discrete time of loop quantum cosmology opens up access to the universe before the big bang. (In the water drop, there is, however, no repulsive force that would prevent the singular split.) What is torn apart is the rubber band of the classical picture, not the real atomic world.

Just as it is difficult, if not impossible, for us to see directly back before the big bang, for a life-form—say, an *E. coli* bacterium— settling in the split water drop, some information about the water remaining in the faucet is irretrievably lost, an effect roughly comparable with cosmic forgetfulness. But at and before the time of splitting off, there was contact between the water in the droplet and in the faucet; certain kinds of information can thus reach the drop. For instance, the chemical composition of the water, such as the relative amount of minerals it contains, does not change while a drop splits off: based on a theory for the splitting off that includes this insight, one can infer the chemical composition of water in the faucet from that in the drop, which is directly accessible and easy to determine.

Given sufficient intelligence on the part of the bacterium, there is also the possibility of indirect reconnaissance. By splitting off, the drop starts to vibrate slightly, in a way observable within the drop itself. With sufficiently precise data and a correct theory, one can calculate backward to the time when the drop split off and infer where it came from. Also in cosmology, one makes use of vibrations—matter released in the early universe and oscillating from the interplay of mutual gravitational attraction and inertia, resounding in cosmic tone colors visible in the heavenly intensity distribution of microwaves, or maybe, in the future, gravitational waves. And here one also ventures to draw conclusions about much earlier times.

FURTHER TEST POSSIBILITIES:
NEVER GIVE UP

In addition to the cosmology of the very early universe, there are further situations in which theory can prove itself in a confrontation with observations. Whether quantum gravity effects are large enough to play a sizable role in these cases remains disputed; and even if they are, consequences for our worldview may be less spectacular. But for observations of the highest sensitivity at the outermost frontiers of research, it is always important to be aware of mutually independent phenomena suggesting a new physical mechanism, as seen for the indications of dark energy—in that case, the distributions of background radiation, galaxy maps, and supernovae. Only then can one be sure that a working model indeed has a chance of applying to the whole universe. In the same way, indications for quantum gravity will be convincing only in a strong combination of independent sources.

BIG BANG NUCLEOSYNTHESIS: COOKING UP THE ELEMENTS

During the hot phase of the big bang, the powerful thrust of the initial, possibly accelerated expansion had not only resulted in a universe of already rather large size, but also, like waves arising from a seaquake, had excited particles out of the vacuum. Space was not empty, but contained matter and energy. There was, however, too much energy for atomic nuclei or matter as we know it to exist. All the separate particles, such as the protons and neutrons forming the nuclei of atoms, just buzzed around unbound. And if perchance one of the busy protons and a neutron had come together in strong, tender bonds to form a common nucleus, they would instantly have been shattered by another energetic particle's intrusion.

Even individual particles were very short-lived, for the high density of matter made it very likely for them to encounter their neme-sis in the form of an antiparticle, each time to disappear completely in mutual annihilation—except for their total energy left in the form of photons. But particles were not eliminated altogether; the pho-tons, when scattering off charged matter, were able to re-create pairs of particles and antiparticles, sending them back on an arduous trip to their next demise.

All this slaughter did not go on without end, for the universe expanded, thinned out, and cooled down. The energy of photons slowly dwindled until it no longer sufficed to create particle-antiparticle pairs. At first, protons and antiprotons, neutrons and antineutrons ceased to emerge from pair creation, while the produc-tion of electrons and their antiparticles, positrons, continued. With a mass of only about half a thousandth of the proton mass, they are created with less energetic effort. Protons and neutrons, still annihi-lating themselves in encounters with leftover antiparticles, thus began to disappear from the universe, leaving behind only their energy.

One of the big puzzles of cosmology and particle physics is why protons and neutrons were able to live until the present day in large numbers, instead of having been completely annihilated in encoun-ters with antimatter.[8] One can be fairly certain that all visible matter is indeed matter and not antimatter: Even between galaxies there is hydrogen gas that, when annihilated at an antigalaxy, would emit clearly visible radiation identifiable by its characteristic energy. To solve the puzzle, one cannot simply assume that matter and antimat-ter are separated from each other, their Armageddon prevented by the peacekeeping force of endless space. A disparity in favor of mat-ter must either have existed from the very beginning, or have come about in the course of time.

But this view, too, brings only troubles. An initial disparity could not be explained but at most be postulated. On the other hand, it is not easy to envision a disparity building up under the known physi-cal laws. (Possible scenarios were first discussed in 1967 by Andrei Sakharov.) There are symmetries between particles and their antipar-ticles, strongly correlating their production and interaction rates. While these symmetries are not perfectly achieved, the degree of vio-lation, known from accelerator experiments, is very small. No con-vincing mechanism has been found yet which would naturally lead to a sufficiently large disparity in favor of matter.

Quantum gravity cannot offer much either, and so we have to accept this fact for now. The universe, then, contained a sea of elementary particles that by and large left one another alone and, except for electrons, remained undisturbed by antimatter: There were electrons and positrons, protons, neutrons, and the ubiquitous neutrinos as well as highly energetic photons. With its expansion, the universe had became so cool by the end of the big bang phase that most of the freely moving particles had lost much of their energy and were at last allowed to settle down by forming stable nuclei. This transition is called *big bang nucleosynthesis*.

Compared to the multitude of chemical elements that can now be found on earth, big bang nucleosynthesis produced only a humble handful—merely a heavy isotope of hydrogen called deuterium (with not only a proton but also a neutron in its nucleus), helium, and a few other light elements such as lithium; but the exact ratios of these elements as they emerged back then are important for the course the universe and its contents subsequently took. The largest fraction came from hydrogen—75 percent—followed by helium with almost all of the remaining 25 percent. Deuterium contributed only about one nucleus out of a hundred thousand others, and lithium just a few nuclei out of ten billion others. The nuclei of all the heavier elements together were created with a mass fraction of just one percent. Nucleosynthesis in particular of the heavy elements continued later on in stars, where it produced the whole of planetary matter; this thread will be taken up again in the section on the first stars in our next chapter. But for the correct element ratios as they can be seen now, the distribution provided by big bang nucleosynthesis was already decisive.

The processes dominating big bang nucleosynthesis are so sensitive that the smallest variations in the ratios of elementary particles and in electromagnetic radiation can lead to strong deviations from observations. This is another opportunity for theories of gravity to prove themselves, because the quantities of particles are determined by the dilution behavior of the expanding universe, and vice versa: The total quantity of different particles influences the expansion rate as a result of their gravitational attraction. From the ratios of elements shortly after the big bang one can thus draw conclusions about the gravitational force active at that time.

In this regime, the repulsive forces of quantum gravity are no longer prevalent, but even the attractive contribution to the force can be subject to small changes compared to general relativity. These changes could upset the fine balance required for successful nucleosynthesis; their size and the underlying theory can be constrained, at least in principle. Such constraints, in turn, can be used to analyze the atomic structure of space-time responsible for the deviations from classical forces. Also here we have some kind of indirect microscope: Even without a direct view on space-time atoms, their properties can be probed by effects on much larger elementary particles. The required calculations, however, are quite complex and are still being developed. According to studies that I recently undertook with Rupam Das and Bob Scherrer, it seems that much more precise observations of the element abundances from nucleosynthesis than are currently available will be required in order to test quantum gravity in interesting ways.

SPACE-TIME AS A CRYSTAL: EMPTY GEMS

Given sufficiently high energies, the recent universe also allows effects that, like a microscope, could resolve the atomic structure of space-time. For light propagating in the universe, space-time itself, if it is indeed of an atomically discrete form, behaves much like a crystal. In an unstructured medium such as a vacuum, light of all wavelengths moves in the same way, in particular with the same speed—that of light, of course. In a medium of atomic nature such as a transparent crystal, by contrast, the various colors are subject to different propagation rules.

Light propagates in a medium by exciting charged particles to oscillations, allowing it to jump from atom to atom and thus propagate through the whole crystal. For light of shorter wavelengths—of higher frequencies closer to the blue than the red end of the visible spectrum—it is more difficult to excite the tardy crystal atoms to vibrations: If the wavelength comes close to the distance between crystal atoms, it gets more and more complicated for it to reach neighboring atoms. The medium appears stiffer than for waves of longer length, and the propagation speed decreases.

A well-known consequence of this so-called dispersion is the splitting of white light into bands of single colors, as in a rainbow or a prism. Changing velocities influence the refraction angles of waves at an edge where a crystal borders on, say, air. With two such non-parallel edges intersecting at a corner—as in a prism—white light, as a mixture of all colors, striking one side, leaves the crystal by the other side split into all its colors.

If space-time is atomic, resembling a crystal, light of different wavelengths no longer travels at a uniform speed even in the absence of matter. When moving through empty space, from a distant star to us, dispersion effects arise: Waves of different frequencies arrive here with different delays even if they were emitted simultaneously by the star. As before, individual corrections compared to propagation in a structureless space-time are very small. But if the star is distant enough, those tiny effects could add up during the long travels of light, and a potentially observable time delay would result.[9]

For a large delay after long travels, the star must be far away, making it more difficult to see. Moreover, one cannot use a uniformly glowing star; that would deprive us of any possibility of determining which parts of the radiation were emitted at the same time. No information about their time delay at arrival would result. What is needed is a faraway eruption of radiation, an explosion of short duration and yet more intense than from supernovae. The short eruption time combined with the required far distance means that we are talking about unimaginably powerful explosions. Fortunately, such events exist in the form of *gamma ray bursts,*[10] exploding in fractions of a second and yet releasing more energy than a thousand suns in their whole lifetime!

Initially, these bursts were thought to be events in our Milky Way; it was difficult to imagine how their great brightness could be caused by a very distant object. But when more and more of them were found over time, they showed a largely isotropic distribution over the whole sky, in no way following the structures of the Milky Way. Gamma ray bursts must thus be objects outside and far from our galaxy.

Moreover, these gamma ray explosions are so energetic that they emit the majority of their radiation not as visible light but as gamma particles—known also from a form of radioactivity. Just like light, these are electromagnetic waves, but of extremely high frequency and short wavelength, endowing them with another advantage for scrutinizing space-time structures: The wavelength is closer to the expected scale of discrete space-time than it would be for visible light, even though the difference from the Planck length remains extreme, and thus corrections to its propagation by the crystalline structure are more pronounced and more easily detectable. (Then again, the frequency should not be too high owing to quantum theoretical effects of the electromagnetic field. Radiation is composed of photons—energy packets carrying the radiation's total energy. With high energies of single packets, and thus high frequency, the total radiation is distributed over few packets. Fewer quanta then hit a detector, and the measurement result is subject to stronger statistical fluctuations, from which the precision suffers.)

In June 2008, the satellite GLAST (the Gamma-ray Large Area Space Telescope, now officially called the Fermi Gamma-ray Space Telescope, but still the old acronym) was launched, dedicated to a precise detection of such eruptions. It has already provided many new insights into the behavior and distribution of these stellar explosions. Whether this satellite's results will be significant for quantum gravity remains to be seen, but it shows good potential. The main purpose of the satellite is, foremost, to provide a better understanding of the origin of these bursts that bring us knowledge from the depth of space. In the meantime, a lack of knowledge of their origin certainly does not prevent us from using their radiation for other purposes (such as testing quantum gravity). One possible explanation for the slightly longer eruptions is a star collapsing to a black hole. Peter Meszaros and Martin Rees, in the early 1990s, analyzed how the collapse energy can be converted into gamma rays, in agreement with many observations. (An alternative explanation for shorter eruptions is the merger of two neutron stars, whose end result could again be a black hole.) Irrespective of whether this is the case, black holes provide a further stage for explorations of quantum gravity.

BLACK HOLES: WORDS OF BLACKNESS

Black holes are plagued by the same singularity problem as the entire cosmos. It isn't the whole universe that collapses here, but only some part of its matter content gathered in a bounded region. Still, in the end waits the breakdown of general relativity in the form of a singularity, grimly reaping not just all of matter but space and time themselves. The repulsive forces of quantum gravity can prevent the total collapse and the singularity, as we will see in more detail in the next chapter. For now, we are interested primarily in the observability of key effects arising in the process.

Black holes are extremely uncommon objects, and so their existence in the universe was long doubted (and occasionally still is). By now, however, they have been clearly confirmed by many observations. Some astronomical objects can, given their compactness, be understood only as black holes. They appear as highly compressed distributions of mass in a small spatial region, with an enormous density compared to all other stellar objects (and yet much smaller than the Planck density). As already seen, gamma ray bursts can possibly be traced back to black holes, but this explanation is not completely confirmed.

The strongest evidence for black holes comes from the center of our own galaxy. Infrared telescopes have allowed us to glance into this cluster of stars for some time, despite the intense light surrounding it. One can even identify single stars and follow their trajectories in this region. It turns out that they revolve around the galactic center in much the same way that planets do around the sun.[11] One deduces the gravitational force acting on them from the orbital size and velocity. The force, in turn, reveals the mass concentrated in the center, and it is enormous: millions of solar masses. But then, no central object such as a star can be identified: The mass is concentrated within too small a region. This gargantuan mass is contained in a region about the size of a single star. If it were a highly compressed star, it should radiate fiercely; but that would make it visible from afar. There is no explanation other than an immense black hole as the center of the galaxy, a tough drillmaster eating its own light and forcing an army of stars, the whole galaxy, to surround it in an elegant march.

Usually, black holes are very heavy, because enough collapsing matter must be present to generate their strong gravitational force. If just a relatively small amount of matter collapses, of about the size of our sun, most of it is ejected before a black hole can form. Since the effects of quantum gravity in a large universe—as in the context of dark energy—are expected to be small, they are also weak in the neighborhood of heavy black holes. They can be large in their center, near the singularity of highest density, but this region is buried deep in the black hole, cut off from the exterior and from observations. While a black hole does not have a surface like a planet or a star, it is surrounded by a bizarre cover called the horizon. Its properties will be seen in more detail in the next chapter; for now we just keep in mind that this veil, for a massive black hole, is far away from the singularity, preventing access to strong quantum gravity effects. For possible observations, we must content ourselves with weak effects in the cover's neighborhood.

Around the cover, there are no strong quantum effects of gravity, but those of matter do arise. The structure of space-time changes dramatically, for here we are at the boundary between the exterior, where falling into the black hole could still be prevented by sufficiently strong forces, and the interior, where this is no longer possible. Once the horizon is traversed, there is no holding back; in the classical world one is irresistibly drawn in to be shattered at the singularity. In the interior, the gravitational force is overwhelming and cannot be compensated for by any other force.

In this behavior, space-time differs from our well-known, familiar cosmic neighborhood and resembles, as it did during inflation, a trembling, shaking ground. As in the cosmology of an accelerated universe, matter is created from the vacuum: quantum fluctuations lead to a particle number differing from zero or, more intuitively, the different balances of forces within and outside the cover can tear apart particle-antiparticle pairs. One member of the pair falls into the black hole; the other one can escape and possibly hit a detector much later on. This phenomenon, discovered by Stephen Hawking in 1973, is known as the Hawking radiation of black holes.

The rate by which radiation disappears from the neighborhood of a black hole is very small for massive black holes, much less even than the weak background of cosmic microwaves. Large black holes will not be discovered by this phenomenon because their Hawking radiation simply drowns in the microwave background. For black holes already identified, the radiation is clearly too weak to show these quantum effects. Smaller black holes do have more intense radiation: For them, the cover lies closer to the central singularity and force differences in their neighborhood are more pronounced. With a sufficiently strong force, the radiation should be more intense than the cosmic background; such black holes lose more energy by Hawking radiation than they gain by absorbing background radiation.

Once Hawking radiation reaches an intensity above the background level, energy loss from the black hole, slow but inescapable, results. The covering horizon moves ever closer to the center, further enhancing the radiation. Slowly piling up the loss of energy, the black hole must either evaporate completely after a finite time, or hope that quantum gravity changes the balance once the cover comes to lie close to the central singularity. This could open up new possibilities for observation, provided that such objects have existed in the recent past of the cosmos and did indeed evaporate. What exactly happens with such a black hole remains unclear; possibilities range from complete evaporation to the disappearance of the cover, making the bare interior visible from the outside. The latter case would manifest itself by an explosion revealing a part of the collapsed matter. (This can only be a fraction, since some energy was already radiated away in the Hawking process.)

Can we really employ this effect for observational tests of quantum gravity? Probably not, as fascinating as it would be. The cosmic background radiation is currently too intense, even though it has already cooled down to -270 degrees Celsius. As weak as this radiation is, the Hawking radiation of most black holes is weaker still. Very small black holes are necessary for evaporation to take place at all; but in this case it would happen so quickly that such black holes must have evaporated long ago. New and powerful telescopes might one day glimpse such events.

As one can see, we do not really live at a good cosmic moment in time, at least if one's aim is to observe Hawking radiation. Going back to the thoughts at the beginning of this chapter, especially in the case of cosmological observations we are obliged to live with what the universe presents us. Later in the history of the universe we might get another chance: The microwave background will be diluted so much as a result of cosmic expansion that even heavy black holes could evaporate, giving rise to entirely new phenomena.

6. BLACK HOLES
COLLAPSE, AND HAPPY ENDINGS?

On the face of the sun its countenance gazes,
Then all of a sudden nothing is there!

—GILGAMESH EPOS

Kruskal"[1] *the capsule is just about to break into pieces. Ever deeper it penetrates the hot dense mash of matter, only barely recognized as a collapsing star. It is unmanned, for no strong human and not even a single cell would be able to sustain these forces. Using modern femtotechnology, it has specifically been made so tiny that the forces yanking at its different parts are minimized to the greatest possible extent. Its nuclei swing around wildly, whipped by the mightiest anger of a foaming space-time. The miniature instruments withstand those conditions only with luck. No chances of retreat or of repair exist now, for long ago the capsule had passed the black hole horizon. Being pulled ever closer to the center, it would be unable to escape even if it had the strongest engine.*

The further the capsule proceeds, the stronger the tidal forces acting on it: Here space-time is curved so much that the gravitational force on one side differs greatly from that on the opposite side, even considering the small size of the tiny capsule. Helplessly the capsule is deformed by the black hole's forces of nature, as happens on a smaller scale with the tides of the oceans and even the continental land masses in the gravitational tug of war between earth and moon. The material forces can withstand this only for a limited time, and soon the capsule will be torn apart. Even its constituents will splinter ever more, before at last dissolving into the smallest particles.

But until then the capsule's heart, the Chiao transformer, will do its valuable work. It is the brainchild of the physicist Raymond Chiao, who once postulated the possibility of transforming electromagnetic waves into gravitational ones. Following this principle, the capsule changes a fraction of the radiation that continuously hits it, sharing its inescapable fate after having fallen through the horizon, to gravitational waves. So it emits gravitational waves of a frequency depending on the deformation of the Chiao transformer, and in this way it encodes the experienced tidal forces into a gravitational wave signal. Moreover, the capsule, when released by the satellite Hawking 3, was put into vibration such that there is a rapidly varying oscillation in addition to the changing deformation. The signal is modulated characteristically, making it easily identifiable should it be caught, some time in the distant future. Thus the tortured capsule sends out a message, rushing ahead in its uncertain travels . . .

A future mission into a black hole might proceed something like this. But as one may suspect, this is quite unlikely, for who would organize and finance such a mission? The capsule would certainly be destroyed in the end—a fate not essentially different from that of many other satellite missions in the solar system. But here, on a mission into the interior of a black hole, not even signals carrying information about measurements can be transmitted to us as observers remaining at a safe distance. Even if the capsule finds enormous treasures, it will reveal nothing.

The only chance to have this mission funded is the action of quantum gravity! Resulting repulsive forces could prevent the total collapse of a black hole into a singularity and possibly cause conditions allowing matter such as the capsule, or at least its signals, to escape. At present, no physicist would seriously propose such a project, but a theoretical trip into black holes is free of such limitations and provides highly interesting insights. Why, in the first place, is it not possible to escape, not even for light? And how would the capsule perceive the central region of the black hole? Is the center a point in space, like a tiny but extremely massive sun one can see ahead, threateningly bright, while inescapably being pulled in?

As we will see, the central singularity of a black hole in general relativity is of an entirely different kind. One cannot see it even if one has already fallen in; one can feel it only by its ever-growing tidal forces. It is invisible because it simply does not yet exist! The singularity forms only in the instant when one is just about to hit it. To visualize this at least to some degree, we have to take a wider view and first look at the formation of black holes and the impossibility of escape from them.

ON THE WAY TO A BLACK HOLE:
PLAYING WITH FIRE

Gravitational collapse leads to denser and denser compressions of concentrated matter resulting from the gravitational attraction between all constituents. For stable stars, it can be slowed down by counteracting forces only in limited ways. Starting with an average star that has spent its initial fuel—the hydrogen concentrated in it after big bang nucleosynthesis, as discussed in the preceding chapter—there are several steps of stabilization, depending on the initial mass.

THE FIRST STARS: BURNING THROUGH THE FUEL

Most visible stars rely on fusion to draw away the energy they radiate in the form of visible light and other waves. As in big bang nucleosynthesis, the two lightest elements—hydrogen and helium—initially take center stage. Hydrogen contains a single proton in its nucleus, constituting the entire nucleus in the majority of hydrogen atoms; but in heavy hydrogen (deuterium), it is accompanied by a neutron. The two particles stay so close to each other that the resulting nucleus is much smaller than the radius of the atom, whose main extension comes from the space occupied by its electron's wave function. There is also the hydrogen species tritium, with two neutrons, but it is unstable, with a short decay time of 4,500 days. The neutron does not have electric charge, and thus the nucleus is equally charged in all three cases; it is completed to a neutral atom by a single electron in orbit around the nucleus. All these different forms, or isotopes, of hydrogen are characterized by the presence of this one electron, the characteristic chemical property of hydrogen. Although the masses change if different numbers of neutrons are present in the nucleus, this does not much affect the willingness of hydrogen to interact with other atoms. Similarly, helium comes in varieties each possessing two protons in the nucleus, and in the stable cases one or two neutrons. An electrically neutral helium atom has two orbital electrons.

Two hydrogen atoms can approach each other, initially not held back because of their electric neutrality. On the contrary, it is energetically favorable for two atoms to form a pair, a cohabitation allowing them to share the electrons more economically. In this way, a chemical bond is formed; the two atoms form a hydrogen molecule consisting of two nuclei and a shared orbital wave function of two electrons. The nuclei are kept at a safe distance, the molecular diameter, because they are both positively charged; this is the hydrogen molecule that one usually deals with in chemistry. On Earth, in the presence of atmospheric oxygen, even this molecule is difficult to keep stably in pure form, since hydrogen and oxygen can, again with energy gain, combine to form water in a heavy reaction. In the early universe, however, no oxygen had formed from the first protons and electrons of the big bang, and hydrogen is safe as the fuel of stars.

But how, then, do stars gain energy if there is no oxygen to burn their hydrogen? At first, after big bang nucleosynthesis and with further cooling of the expanding universe, wide hydrogen clouds formed. Like cosmic background radiation, which tells us about those times, and like all matter formed during nucleosynthesis, these clouds are initially nearly homogeneous. But since they are not perfectly homogeneous, more concentrated centers form where the mass density has a small surplus compared to the neighborhood. Even such a tiny mass surplus pulls in, by gravitational attraction, yet more hydrogen from the surroundings, further increasing the density. This is a slow process, but time is something the early universe has in abundance, so some regions can become very dense. As one can see, gravity shows not only those sadistic tendencies culminating in singularities, but also capitalistic ones: Regions already rich in mass become further enriched.

Hydrogen is compressed more and more so that, if gravity had its way, even the nuclei in a hydrogen molecule would approach each other. But they fight back, for they are equally charged and repel each other. An initial balance of forces arises, with electric repulsion counteracting gravitational attraction. This stability is the reason for the existence of planets such as Earth, realized in this case not for hydrogen but for the atoms in Earth's core. Earth and other planets are stable to a high degree because they are not dense enough for gravity to overpower electrical forces between nuclei. There is, moreover, not much matter in their neighborhood, and so their masses cannot increase by further infall (disregarding small contributions such as meteorites). In the much less expanded early universe, this was entirely different: Around the density centers there was still plenty of hydrogen to reinforce the centers even more. Their density rose—so far that gravity pushed electric repulsion into a corner and made the hydrogen nuclei approach each other ever more closely.

That cannot go on for long. When two hydrogen atoms come too close, another reaction happens: Two nearby but separate protons require more energy than a single deuteron (a very close combination of a proton and a neutron; binding an electron, a deuteron forms an electrically neutral deuterium atom). These two particles can come much closer than two protons because the neutral neutron is not repelled electrically. Moreover, the deuteron combination of a proton and a neutron turns out to be stable, with the two particles attracted to each other and bound by the strong nuclear force, provided they are close enough.

This force, fully independent of the electromagnetic force and gravity, acts only over small distances, at least in comparison with common length scales; compared to the Planck length, even these distances are huge. It becomes active only when two protons approach sufficiently closely, as it happened in the early universe— either by random central hits of protons rapidly flying about in the hot phase of big bang nucleosynthesis, or systematically in early post–big bang compressions due to gravitational attraction. One of the protons is then first transformed into a neutron and a positron, the electron's antiparticle, and the other proton binds with the neutron to form a deuteron. Here one of the extremely weakly interacting neutrinos also arises, immediately fleeing the scene. The positron and one of the orbit electrons soon annihilates, and so from two light hydrogen atoms, with their two protons and electrons, a single deuterium atom with one proton, one neutron, and a remaining electron is formed—along with an amount of free energy, as the basis for the star's shining.

Thus hydrogen can be condensed more than would be possible for simple protons, for a single nucleus takes up less space than two separate ones. But gravity has no mercy. Deuterons must approach each other, as well as the remaining common hydrogen nuclei, more and more, until at last further reactions happen. For instance, two deuterons have exactly the same combination of particles as one helium nucleus. And indeed, when they come close they can combine to form helium, again gaining energy thanks to the binding nuclear force. Like two protons to a deuteron, so two deuterons fuse to a helium nucleus—a process currently raising high technological hopes as a possible source of energy on Earth, as the fuel, hydrogen, would be nearly undepletable, and the process itself would be very clean; the "waste" would, after all, be helium.

In stars, fusion is easily possible because gravity compresses hydrogen without further ado, and the high pressure makes the nuclei approach each other closely enough to make the reaction happen. Much more energy is released than in the formation of deuterons, exactly the bulk of energy radiated away by stars. In the stars' interiors, this leads to an increase of temperature, and the heated matter is subject to a strong pressure able to withstand gravitational collapse. Thus stars like our sun are stable for long periods of time—a lucky break for our survival.

When they grow old, stars slowly use up their hydrogen, including the deuterium. At some point, additional nuclear reactions of helium occur, since an approaching hydrogen or another helium nucleus can add protons and neutrons. Heavier elements form in this way, such as carbon and oxygen, important for life on Earth, but also silicon and iron, which form the largest part of the planet. Energy is released in these reactions, contributing to the radiation of stars; but it is no longer as much energy as in the star's youth. The most stable nuclei turn out to be iron and nickel, where fusion stops; elements heavier than those are produced in stars only in minor quantities. (Most heavy elements as they can naturally be found on Earth form in violent stellar explosions rather than in normal fusion burning, adding protons and neutrons to iron in rapid fire.)

By these long-lasting processes repeated uncountable times, keeping the oven hot for billions of years, the first generation of stars cooked up a large variety of elements, much larger than was possible during the first fiery minutes of big bang nucleosynthesis. Some stars—exhausted by their mass enrichments—exploded in supernovae when they had used up all their nuclear fuel, releasing the elements into space. Long afterward, the stellar dust again gathered in clouds, just like the initial hydrogen, and started once more to form density centers. From one of these dense regions, in a later generation after the first stars, our own solar system and its planets, including Earth, were formed.

WHITE DWARFS: COLD AND COMPACT

When its nuclear fuel is nearly depleted, a star can no longer shine as intensely as during helium fusion. It will initially swell and become a so-called red giant, glowing for some more time. The sun will share this fate in about five billion years, unfortunately swallowing Earth. But when all the fuel is gone, the old dilemma is faced again: Gravity insatiably compresses the star, much more intensely than before the state of the red giant. What counterforce can now become its equal to make a stable object possible?

A light star will, after using up all nuclear fuel, simply cool down and shrink somewhat, until the material pressure suffices for stabilization as in a planet. But this is not possible for heavy stars; there is simply too much mass pushing on the interior of the faded star. Such high pressures cannot easily be produced in the laboratory, and so the behavior of matter in such a state is not completely understood. For a certain range, however, more familiar physics suffices to understand the further collapse of a star.

At first, quantum mechanics rushes to help, as it already did to stabilize a hydrogen atom. It leads to a new but unfortunately extremely counterintuitive force able to hold its own against gravity. As we have seen, electric repulsion has long failed at such high densities. But there is another reason for two particles of the same kind, such as two protons or two electrons or even two electrically uncharged neutrons, to stay away from each other. This was first realized by Wolfgang Pauli and formulated as a principle: Pauli's exclusion principle, which in 1945 gained him the Nobel Prize in Physics. The use of the term "principle" indicates that this, like so much else in quantum mechanics, has a rather abstract origin. But it is a very real phenomenon, as shown not only by the existence of white dwarfs in the universe but also by a multitude of experiments on Earth.

What we perceive as pressure in ordinary matter, as well as in common stars, is caused by the impact of atoms or molecules on one another. Higher pressure arises from more vigorous pushing of atoms or molecules on surrounding matter, which could result from a raised temperature. But without nuclear fuel in old stars, the ability to increase pressure by heating up is precluded, and the star is doomed to compress under the influence of gravity. The nuclei of atoms, now belonging to heavy and no longer fusable elements, come very close to each other, and their electron wave functions overlap. Once the nuclei are sufficiently close, electrons can easily switch places from the orbit of one atom to a neighboring one; they can no longer be attributed to a unique atom, and instead move much farther away. In this case, one speaks of the formation of an electron gas pervading the dense matter. Electron gas arises at high pressures, pushing atoms close to each other, but it is also realized under natural conditions in metals, where it is responsible for electric conductivity.

In an electron gas, there are no impacts between electrons in the usual sense. Instead, the electron gas resists too strong a compression by behaving according to Pauli's exclusion principle; the particles are held apart solely by quantum mechanics. For particles such as electrons, protons, and neutrons, it turns out that two wave functions of the same type—for example, two electron wave functions—cannot occupy the same position. Imagining particles in the classical picture as pointlike creates no problem, because a point does not require space. However, the wave function in quantum mechanics must be spatially spread out, and so the exclusion principle, which dictates that two wave functions cannot be at the same spot, means that quantum mechanical electrons must repel each other. This happens when the wave functions come too close, usually at very small distances realized only by high compression.

Such a force requires a rather advanced collapse of matter until it can act. And so the resulting object stabilized by this strange quantum mechanical power is very small, a white dwarf: a compact object created by gravitational compression after almost all nuclear fuel is used up. Nuclear reactions no longer take place in the interior, and a high temperature is now unnecessary for stabilization. Strange as they may appear, these objects are rather common, with an average distance from one another of about ten light-years in the Milky Way, compared to a total galactic diameter of 100,000 light-years. Their mass range is rather small by the standards of usual stars: White dwarfs have a minimal mass of a thousandth of the mass of the sun, and they reach masses up to one and a half times the solar mass. Owing to their strong compression, they are nonetheless very small, with radii all below that of the planet Jupiter.

NEUTRON STARS: THE LAST UPRISING

Is a white dwarf finally the stable answer to gravity? As the upper limit of one and a half times the solar mass suggests, this is not the case. Nuclear reactions render Pauli's exclusion principle inapplicable if the mass of a dying star is above that maximum for white dwarfs, named the Chandrasekhar limit after Subrahmanyan Chandrasekhar, who was awarded the 1983 Nobel Prize in Physics. When a large mass of matter is compressed in a star, electrons and protons of the elements come very near. (The exclusion principle does not hold for different types of particles.) If they are sufficiently close, a reaction occurs similar to that of two protons fusing. In a somewhat reversed order, this process is also observed in beta decay of radioactive nuclei, in which a neutron decays into a proton. In our case, however, an electron and a proton react and disappear, forming in their place a neutron together with a neutrino, which, because of its weak interaction with matter, can immediately leave the star. Neutron stars owe their existence to this ultimate uprising of matter against gravity.

Under normal circumstances, this reverse beta decay is in principle possible, but the neutron would decay back to a proton and an electron (plus antineutrino) just after a mere quarter of an hour. The neutron is heavier than the other two particles taken together and thus requires more energy to be stabilized. In a very dense star, gravitational collapse provides exactly the energy necessary to stabilize a neutron. And not just one neutron is stabilized—all protons and electrons in the star's interior can easily combine to become neutrons. Again, one particle, the neutron, takes up less space than a proton and an electron together; thus, after the reaction, the star can collapse further and release energy.

This happens explosively. All of a sudden, the star's central region collapses, with an emission of a huge amount of energy accompanied by a flash of the new neutrinos, which is difficult to detect but highly energetic. Large fractions of the star's exterior regions, mere pawns in the violent game of inner processes, are tossed out into space when the sudden impact, brought about by the outer matter collapsing onto the newly formed hard neutron core, causes a shock wave traveling outward. Such explosions as the grandiose departures at the end of an active star's lifetime are widely visible as supernovae. More precisely, these are supernovae of type II (or Ic if no hydrogen is left in the outer shells). They are to be distinguished from supernovae Ia, as encountered in the previous chapter, by virtue of their important role in modern cosmology; the latter are caused by a different kind of thermonuclear explosion of white dwarfs.

What is left in the core is matter even more uncommon than that in a white dwarf: almost pure neutrons. Here, no electric forces are acting, because all particles are neutral. The only stabilizing force is the quantum mechanical repulsion of neutrons, again based on Pauli's principle, which does not care about the absent electric charge. Compared to a white dwarf, a neutron star is even denser, for from two charged particles, an electron and a proton, a single neutral neutron was formed. The wave functions of all neutrons thus take up less space than those of electrons and protons before. They can be compressed much more, allowing neutron star matter so dense that one cubic meter of it weighs more than a trillion tons. Similar to white dwarfs, whole neutron stars can be up to twice as heavy as the sun; but owing to their high density they only occupy the space of a sphere with up to ten kilometers of radius. At the surface of a neutron star, the gravitational force can be so strong that sometimes even nuclei are deformed by tidal forces.

Neutron stars do not emit much energy after the supernova dwindles down, but one can often find them by their gravitational attraction alone. It happens quite frequently that a neutron star comes sufficiently close to an ordinary one to remain in mutual orbital motion. As a result of the joint motion of both partners around their center of mass, the visible star shows a stagger, allowing a mass estimate of the neutron star to be inferred. Neutron stars also play a large role in pulsars and in some experimental tests of general relativity, as previously described. These objects are clearly confirmed in the cosmos, and they are stable for long intervals.

Does the hard matter of pure neutrons confine gravity within its bounds? No, for here, too, there is a limit to the quantum mechanical force, as again indicated by the presence of a maximal mass of neutron stars. If the neutron star is heavy enough—about twice as massive as the sun—Pauli's repulsion is overpowered. The precise value of the upper limit is theoretically much less determined than for white dwarfs; neutron matter is, after all, known much less well than the dense matter of a white dwarf. But luckily, one can perform calculations in general relativity that clearly and inescapably show the breakdown of all conceivable forces in matter of sufficiently high pressure, or for sufficiently heavy stars. The generality of this statement, independent of the precise form of matter, once again underlines the elegance of relativity.

When this last rope ruptures, there is no holding back. Matter now collapses unhindered as a result of the unchecked force of gravity. No known physical laws of matter provide further counterforces to play a role here. One could expect hitherto unknown physical laws in such highly compressed and energetic matter, bringing about new stable stellar objects. But nothing of this kind has been found, and nothing that matter could conceive would stand a chance against general relativity's untethered gravity. What one sees instead, in the actual universe as well as in mathematical solutions of general relativity, are black holes: the final stages of the total collapse of matter.

THE CENTRAL SINGULARITY:
A DREADFUL POINT IN TIME

No one at all sees Death,
No one at all sees the face of Death,
No one at all hears the voice of Death,
Death so savage, who hacks men down.

—GILGAMESH EPOS

The concept of black holes, as it initially arose from solutions of general relativity, can convincingly describe many properties known astrophysically from very dense regions. There is no avoiding them; extremely compact objects far heavier than the maximally allowed masses of neutron stars do exist in the universe. In the center of our own galaxy, for instance, there is a compact mass, called Sagittarius A*, as heavy as about three million suns; and yet its size cannot be measured from Earth. Such a beast can only be a gigantic black hole.

Taking the complete space-time literally, as it results from the classical equations of general relativity, all matter in a black hole collapses to a single point. As at the big bang, we encounter a singularity of infinite density, making the theory's equations lose their validity. Thus, a wider theory must be developed so as to remain valid and precisely show us the final stage of gravitational collapse. A candidate for such a theory is, of course, again quantum gravity. But even before the total collapse, black holes of general relativity show properties different from those of neutron stars in curious ways. For instance, the singularity is not a point in space, as alluded to above, but a point in time. These properties, so important for an understanding of black holes, are our next subject.

HORIZONS: POINT OF NO RETURN

When matter has collapsed completely and no force counteracting gravity can keep the elementary particles apart, does a black hole then appear as a single, infinitely dense point in space? The answer is no, for two reasons: In the final stage, matter has completely collapsed, but the curiosities of space-time transformability in general relativity do not make this happen at a point in space. Instead, one could better describe it as a "point in time," as will be discussed below, but even this does not reflect all properties. Moreover, one cannot see this point in time at all from outside a black hole; the singularity of matter collapse is shrouded by a horizon far away from it. Only the properties of this horizon and its surroundings reveal the existence of a black hole from outside. No direct insight by observations yet exists; but in about ten years the satellite observatory Constellation-X is scheduled to explore the space-time neighborhood of a horizon around a massive black hole by recording emitted X-rays.

All these properties were already contained in the solution found by Karl Schwarzschild for nonrotating black holes in 1916 — just one year after the first publications of general relativity. However, the singularity, as well as the existence of a horizon, were long misunderstood, fully decoded only in the 1960s with progress in the geometric understanding of general space-time properties. Einstein himself never believed in the reality of singularities threatening his theory, from which it must, if it is taken as fundamental, perish. Instead, Einstein, like most physicists of his time, thought that a singularity arises only mathematically through the assumed strong symmetry, but that it would disappear in more realistic solutions. (However, Einstein repeatedly emphasized that he did not consider general relativity to be fundamental, but instead expected necessary extensions arising from quantum theory and the atomic form of matter.)

In an exactly spherical matter distribution, as described by Schwarzschild's solution, collapse can happen only centrally toward a single point. There, all matter will gather and quickly lead to infinitely high densities. One could assume deviations from the exact spherical symmetry of the initial star, as always occurs in realistic cases, to result in a collapse into a more extended region; the end product might be highly concentrated but nonsingular. But this is not the case. Singularities, as analyzed in what follows, always occur, and they must be considered a real threat to the theory of general relativity.[2]

Some of the geometric techniques used for an analysis of such bizarre objects, where the intense curvature distorts not only space-time but also our imagination, have an intuitive basis. They go back to a construction introduced by Roger Penrose called *conformal completion of space-time.* "Completion" here means that all of space-time, extending all the way to infinity, is mapped to a much tidier finite region. "Conformal" means that the mapping, while squeezing space-time into that finite region, does not deform it too strongly and preserves geometric shapes, especially the size of angles. This is particularly important because angles in space-time, as already seen, are to be understood not just spatially but also in a spatial-temporal sense, as velocities. Velocities are respected, too—in particular, that of light. And the speed of light as the absolute top speed in space-time traffic has a special meaning; this important physical law is thus respected by the conformal mapping.

In the case of Schwarzschild's rotationally symmetrical solution, one fortunately need consider just two of the four space-time dimensions: the radius and time. Rotational symmetry means that properties of space-time, like those of a perfect sphere, do not depend on the orientation angle around the center. The two angles required to specify one's orientation in space can be disregarded without overlooking essential properties. One can visualize such space-times by two-dimensional, rather than four-dimensional, diagrams, in particular in combination with the conformal completion, called Penrose diagrams. (In the case of the Schwarzschild solution, this diagram is also referred to as the Kruskal diagram, after Martin Kruskal, who introduced suitable coordinates in 1960.)

The intuitive basis of a Penrose diagram is the following: As often is the case in astrophysics, a concrete investigation of black holes is a scattering problem. A black hole does not emit its own light, and the aura of Hawking radiation emerging from its neighborhood is usually too weak to permit observations. Instead, one can recognize a black hole by light emitted from other stars that passes near it and is deflected into our observational instruments (or away from them). As with the moon, one can "see" a black hole and determine its properties by the scattered radiation, even though this comes with additional complications arising from the great distance and the special properties of scattering. (The black hole, after all, does not have a distinct surface but only a horizon.)

On the basis of this scattering problem we can visualize black holes in two-dimensional representations. To that end, we draw light rays in a space-time diagram (figure 20), containing the radius and time. Time in relativity is usually drawn upward; we thus choose the vertical direction as that of time. The radius can then change to the left or the right, depending on whether we approach the center (leftward) or move away (rightward). But the radius, in contrast to time, is always positive: It tells us the distance to the center, but not the direction. We have to add a boundary to our diagram, in the form of a vertical line. On this line, the radius takes the value zero; we are at all times in the center of spherical symmetry.

What happens when we approach this line along a physical trajectory? This is not a boundary of space-time analogous to the big bang singularity; at this place nothing unusual happens. The line merely represents a special position, since it sits in the center of symmetry. When we, or light rays, approach the center, it is simply penetrated,

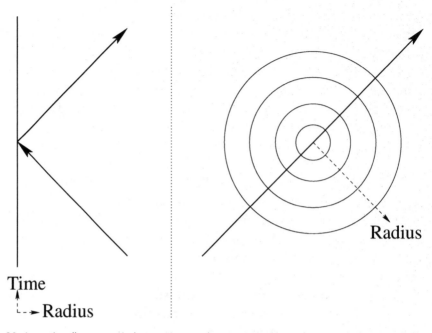

Time
Radius

20. A rotationally symmetrical space-time can be represented by a plane on which time and the radial distance from a center function as coordinates, bounded as a result of the positivity of the radius. Light rays run along straight lines, inclined by 45 degrees. When they hit the boundary, they are reflected in this representation, for they traverse the center in such a way that their distance to it first shrinks, then vanishes, and finally grows back. Left: Space-time diagram with the boundary at left as the rotational center in time changing vertically, and with the trajectory of a light ray moving through the center. Right: Spatial diagram of the same situation, with the rotational center at midpoint and the trajectory of a light ray at different times. The radius—the distance of light from the center—first decreases, then vanishes at the center (the left boundary of the space-time diagram), and finally grows after passing through the center.

as in figure 20. The center is not a physical object and thus cannot influence matter or light. When something passes through, just the radius as the distance to that point changes: It first shrinks, then vanishes, and finally grows back. The radius never becomes negative as required, but it flips its direction of change from decreasing to increasing. In the two-dimensional space-time diagram, this appears as a "reflection" of the light ray at the vertical line; coming from the right, it runs back to the right after having touched the boundary. Nevertheless, there is no material such as a mirror. The "reflection" is merely a consequence of our intuitive two-dimensional drawing of actually four-dimensional events by ignoring the spatial angles.

With this, we already have the first ingredients of Penrose diagrams: a vertical line symbolizing the behavior in time of the center, and light rays. Since light in empty space always has the same velocity, as a space-time angle unchanged by the conformal mapping, we fix the direction of light rays to be always 45 degrees, to the left or to the right depending on whether light moves from large radii to small ones or the other way around. One can imagine these rays as coming from infinity and moving back to it after having traversed the center. On the right-hand side, no boundary lines are necessary. However, boundedness comes from the mentioned "completion" of space-time to a finite region, and so one often draws an empty space-time as an isosceles triangle, with the vertical line of the center as one boundary and the finitized origins of incoming light rays and the targets of outgoing light rays as the two other sides, as in figure 21.

The result is a complete two-dimensional diagram of a space-time where only light rays propagate. (Having disregarded spatial angles, we can only draw light rays moving through the center in our diagram. This is sufficient for an understanding of space-time and its scattering properties.) Empty space, except for light rays, is certainly not very spectacular; but one can easily extend it by further elements to describe the spherical bodies of astrophysics. Then, the 45-degree lines show the scattering problem that represents astronomical observations.

Perhaps surprisingly, a compact object in the center, for instance the moon, does not present itself much differently from empty

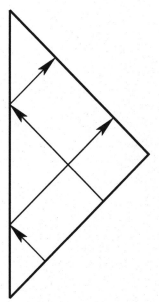

21. Complete space-time diagram with two propagating light rays. Coming from the past, corresponding to an arbitrary distance away from the center to the bottom right, they approach the left vertical line, traverse the center, and move away from it into the future. This illustrates a scattering problem, with light shining from a light source at the bottom right and being measured by detectors at the top right. Both the light source and the detector move upward in time along the boundaries. The diagram visualizes a whole history of events of emission and detection. That the lines of sources and detectors intersect with the vertical boundary, seemingly emerging from or falling into the center, is an artifact of the completion procedure. As parallel lines in perspective drawings seem to intersect at their vanishing points, the points at which the boundary lines of a conformal completion intersect do not exist in reality but lie at infinity.

space. The object has a certain radius, so we must draw its trajectory in time somewhere to the right of the left boundary (see figure 22). The lunar radius is constant; the surface should have a fixed distance from the center at the left boundary. But distances, unlike angles, are not respected by the conformal mapping, and thus the surface is generally represented by a curved line in conformal diagrams. This avoids having the surface intersect the two borders to the right (except at the corners), which are supposed to be reached only by light rays.

At the curved line, as with the surface of the moon, some part of the light falling in from the right is absorbed, and the rest is now reflected in a truly physical sense. The only difference from empty space is that some light rays stop—something we can ignore since lost rays become irrelevant for the scattering problem—while the rest are simply sent back to the right at a different position. From this viewpoint, all compact objects in astrophysics can visually hardly be distinguished from empty space, as shown by a compari-

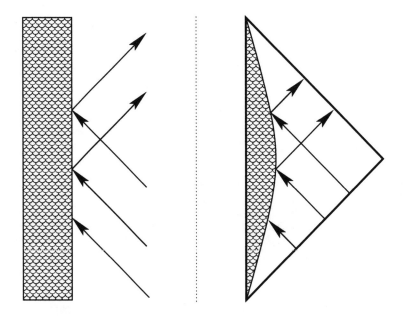

22. Space-time diagram of the moon with three light rays, the first being absorbed, the others reflected. On the right, the whole space-time in its completion is shown. Borders to the right correspond to infinite radii in the past and in the future, respectively, as the origin or target of light incident from afar.

son of figure 21 with the right-hand side of figure 22. Such a broad view may seem a disadvantage, given that our observations of all these objects show important differences. But the similarity of those images—in sharp contrast to that of a black hole, to which we turn now—illustrates in an impressive manner the black hole's special character and its strong effects on space and time.

Significant things happen when we consider a black hole. Bringing these aspects out clearly is the true strength of Penrose diagrams. One could expect the vertical line of the center to become an actual boundary of space-time: the central singularity. This could be one possibility, indeed realized for some solutions of general relativity. But those solutions play hardly any role in astrophysics, for a black hole as it arises in the gravitational collapse of a star leads to a different diagram. As a mathematical analysis shows, the singularity does continue the vertical line of the center, but it does so not in a vertical but, as in figure 23, in a horizontal direction! It is not timelike—not a fixed point at which time would change—but spacelike, a part of space at a fixed time. After all, the vertical position, determining time, does not change along the singularity. For this reason, our initial expectation that the singularity is a point in space was incorrect. Such a point in space, visible at all times, would be vertical in our diagram, just like the center line. Instead, being horizontal, the singular-

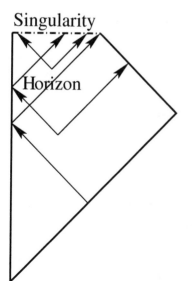

Singularity

Horizon

23. Space-time diagram of a black hole, with the dash-dotted singularity as additional boundary. Not all light rays can now escape to a safe distance, to the upper right boundary. There is one light ray (drawn as coming in from the bottom right) that can just barely escape, marking the horizon by its final segment from the center line. Light rays starting in any direction from within the horizon (in the diagram, above the final segment of the ray that barely escapes) always reach the singularity, even if they move away from the rotational center. Light rays starting outside the horizon and moving away from the center do escape to a safe distance.

ity occupies several spatial points, but it does so at a fixed time: at a fixed point in time.

Collapse into a point is expected to be drawn extended in a space-time picture rather than pointlike, because the point persists in time. The singularity as the end state of collapse is indeed an extended line in the Penrose diagram—but since the process is combined with a space-time transformation deeply inside the black hole, the extension shows itself in space and the pointlike nature in time, and not the other way around.

In general relativity, the new, horizontal line really does represent a boundary: Space-time curvature here is infinitely high, and so is the density of infalling matter that led to this black hole. Here is the point (in time) where matter completely collapses. The equations of general relativity can tell us something about the form of the singularity—namely, that this is a point in time—but they do not lead us beyond it. One cannot use them to explore the possibility of space-time above the horizontal line, since the infinities make all equations lose their mathematical sense. For the collapsing matter, or subsequently infalling matter including unfortunate or sacrificial observers, one can only say that they will reach the boundary after some finite time and then, within this theory, cease to exist. Whether this limit of theoretical existence is also a real limit of the world is, however, a different question that is answerable only with a more general theory such as, perhaps, quantum gravity.

Owing to the singularity's point-in-time nature, it is not visible as a bright spot to an observer approaching it. Were this the case, light would have to start from that point and reach the observer. But all light would travel from the singularity upward in the diagram, into the region where general relativity does not provide us with any space-time to propagate light. The singularity appears entirely different from all other astrophysical objects, and not only because of its extremely high density. Its spacelike nature means that it forms only at the exact moment when an observer falls in!

Although this may appear sneaky, the singularity does make itself known earlier. In its neighborhood—understood in the temporal sense, thus before it forms—space-time curvature is very large, even though it becomes infinite only at the singularity itself. Strong curvature, according to the descriptions in chapter 1, means that the strength and direction of the gravitational force vary greatly even at nearby points. Every extended object, such as a space capsule, is subject to stretching forces when its sides experience different forces tugging its parts in opposite directions. Such forces are called *tidal forces,* by analogy with the well-known, if weaker, effect on Earth, where space-time curvature caused by the joint gravity of the earth and the moon raises the tides. From these forces, one would notice the singularity long before one fell in; objects would be torn apart early in their approach.

Dreadful is the appearance of the singularity for an observer who has already fallen into the black hole and can no longer escape. How is the black hole experienced by an observer far outside? Here, it is important that the space-time portion containing a black hole is divided into two different regions. There is the triangular part underneath the horizontal singularity line in figure 23 from which nothing can escape, and there is the rest of space-time where we would stand as external observers. Since not even light can escape — it can, after all, only move along the 45-degree lines in the interior, all ending at the singularity — one must consider the dividing line as the boundary of the black hole. This is the storied horizon.

The horizon has several important properties. First, it is, except for its endpoint in the distant future, rather far away from the singularity, for heavy black holes, very far away. A black hole of one million solar masses, as it can easily occur in galactic centers, would be as large as the sun. If the total solar mass were to collapse to a black hole, the distance from the horizon to the singularity — the so-called Schwarzschild radius — would be one-millionth of the current solar radius of about 700,000 kilometers. For common stars, the ratio of their current radius to their Schwarzschild radius is similar. For white dwarfs, the Schwarzschild radius is about one ten-thousandth of their radius, and thus comparably higher than for the sun, as a sign of their stronger compression.

Extremely dense neutron stars have a radius just above the Schwarzschild radius determined by their mass. They stand at the brink of collapse into a black hole; were they just slightly denser, their surface would come to lie within the horizon and then be pulled into the singularity. For even heavier black holes, their radius increases proportionally to the mass. In such cases, the horizon is so far outside the high curvature region near the singularity that the space-time curvature around the horizon is quite weak. Quantum gravity is not usually required in this context to understand all the properties—but sometimes one needs the quantum theory of matter, as in the Hawking effect.

Low curvature also means that one would initially not notice one's own passage through the horizon, justifying the use of this term. As a horizon on Earth appears as a distinct border but then, on approach and during passage, turns out to be an immaterial perception, traversing a black hole horizon is initially not accompanied by any threatening or foreboding signs. In the near neighborhood of an infalling observer, space-time appears no different than the exterior space. An outside observer would, however, notice the horizon because signals emitted by the infalling observer will need more and more time to reach a stationary observer outside. Time, after all, proceeds more slowly in a region of strong gravity than it does under weak gravity; time for an observer approaching the horizon passes more and more slowly compared to time as experienced by one staying at a fixed distance. If the infalling one regularly sends out signals, the exterior one can determine the progress and will notice when the horizon passage is imminent.

This behavior is relevant not only for an imaginary experiment in the vicinity of a black hole, but also for astronomical observations. Light is nothing but such a regularly emitted signal. If stars or hot, glowing gas exist in the neighborhood of a horizon, as is the case in the center of the Milky Way, one can notice delays in the periodicity in the form of changing colors: The delay implies a reduction of the frequency, and so one speaks of *redshift* (red light being located at the lower-frequency end of visible light). Such effects are indeed planned to be made use of in future observations to measure the neighborhoods of black hole horizons.

The horizon and its meaning have been understood only rather recently, long after Schwarzschild found his mathematical solution. Grasping its meaning is made difficult by the fact that infinities arise even at the horizon but turn out not to be disastrous for space time. What diverges here is the redshift as it is measured by an exterior observer: The periodic signal of a source falling in, when it reaches the horizon shortly before disappearing, is infinitely redshifted. Viewed from the outside, such a signal no longer shows any temporal change. While outside, far from the black hole and under weak gravity, time progresses normally, processes near the horizon appear delayed. Light itself is influenced near the horizon, largely disappearing from the visible range of the electromagnetic spectrum when it reaches an exterior observer and so, as it were, fading. Here, infinities arise only in the perceptions of an observer; they do not refer to direct physical properties of a material or space-time object. Such infinities are acceptable, but they do require care.

All this gives rise to difficulties when one attempts to describe the whole of space-time by coordinates, similarly to the use of latitude and longitude as a standardized specification of positions on Earth. In curved spaces, it is not always easy to choose globally defined coordinates; even on the spherical space of Earth's surface there are, strictly speaking, problems. At the poles, after all, geographical longitude loses meaning: all longitude lines intersect. Which longitude should one then attribute to the poles? Such points, where some of the chosen coordinates lose their meaning, are called *coordinate singularities.* Mathematically they appear as singular, and they can, if one is not careful, lead to infinities in calculations. But physically, nothing unreasonable is happening at such a point. They are just coordinate singularities, not strict, unqualified singularities.

By the way, the North Pole is often, and misleadingly, used as a comparison with a physical singularity, leading some to assert that it is just as meaningless to speak of "before the big bang" (or, in our present context, "beyond the black hole") as it is meaningless to speak of "above the North Pole" (or "north of the North Pole") if one is tied to Earth's surface. However, this objection is misleading, because the North Pole is just a coordinate singularity, while the big bang singularity and that of a black hole are real, physical ones: There, curvature grows unboundedly, and with it perceptible properties such as temperature or tidal forces. There, one is literally torn apart; at the North Pole one would at worst freeze to death.[3] (And even that happens only because the North Pole has arbitrarily been located in the Arctic, or because one has used Earth's rotation axis for orientation when defining geographical latitude and longitude.)

A simple example for a coordinate singularity can be found on a circle. A single angle suffices as coordinate, but there is always a point where the angle must jump from 360 degrees back to zero degrees. This happens because the circle is cyclic, while we must use nonperiodic numbers as coordinates. It comes to the abrupt jump of the coordinate — a coordinate singularity but nothing more. An ant running along the circle would just move on after each lap, entirely unimpressed by the jump of the angle. The same happens with a calendar: The circle here is the motion of Earth around the sun, and the coordinates—months, days, hours—are just chosen in a more complicated way than a single angle, which would actually have sufficed. Here, too, a jump occurs, namely the one from 2400 hours on December 31 to zero hours on January 1. New Year's Eve is nothing but a coordinate singularity—such a meaningless phenomenon that it seems questionable, certainly from a mathematical standpoint and surely from more practical ones as well, to squander large sums of money on this occasion.

A further important property of the horizon can already be seen from some of the preceding remarks: The horizon is not a surface in the usual sense, not a spatial surface surrounding a voluminous object. After all, the horizon in a Penrose diagram of figure 23 is defined by a 45-degree line, a light ray! No massive object, nor the surface of one, can stay on this line in space-time. It is just as impossible to move up to that line—to land on a surface, as it were—and then move back into the exterior. Once at the horizon, there is no escape: As a material object, one falls into the interior, and soon thereafter into the singularity. Even light can at best linger forever on the boundary and just avoid the crash into the singularity, but it can never get back to the exterior—at least according to classical general relativity.

NAKED SINGULARITIES AND COSMIC CENSORSHIP:
LAW AND DECENCY

As shown in figure 23, the singularity of a black hole, formed by collapse, is a point in time, not in space: a horizontal rather than vertical line in the Penrose diagram. There are other solutions of general relativity, rotationally symmetrical as well, where already the left boundary of the diagram, or at least a part of it, is singular. It can then happen that a part of this singular piece is not covered by a horizon; sometimes there is no shrouding horizon at all in spite of the singularity, as shown by the example in figure 24. This possibility of so-called *naked singularities* has upset some relativists so much that, hoping that general relativity is "modest" enough, they introduced the concept of *cosmic censorship.* Such disrobed singularities are supposed to arise from gravitational collapse only under conditions so special that they play no role in the universe, making them only mathematical possibilities far removed from reality.

Of course, this conjecture has a serious basis: If an uncovered singularity exists, light and much more can reach us from there. From the viewpoint of general relativity, which fails at the singularity and cannot tell us anything about it, literally anything can happen. But when whatever is going on there can influence the rest of space-time, science loses all predictivity.

Something similar, strictly speaking, occurs at the big bang singularity. But here one is so used to the notion of a "beginning," laying the basis for our existence, that one accepts, in fact expects from it, a strong influence on the world. That would be required for an interpretation as a beginning; for when nothing existed before but something does afterward, the singularity must simply leave an extremely dominant impression on the rest of space-time. This transition from nothing to something cannot be explained by relativity, but once it is accepted, at least the further development is computable. The big

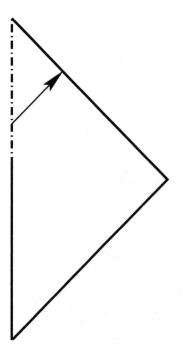

24. Space-time of a naked singularity. While the solid part of the vertical line simply corresponds to a center, as before, an infinitely high matter density is realized in the dash-dotted part. In general relativity, this part can be represented only as an anarchic-singular boundary of space-time. From every one of its points, a light ray can escape to the top right boundary; there is thus no horizon covering the singularity. At the singularity, physical laws break down. No control exists on what happens there. Messages, implications, and instigations of the anarchy can reach outside, rendering all predictions impossible.

bang may be undefined by general relativity, but it happened in the past and is already over. The revolution of a naked singularity, by contrast, lives on, possibly forever. At each moment, the unpredictable can hit us, a far less acceptable violation of the rules of determinism. Differences between the types of singularities show up in the Penrose diagram: The big bang singularity in the space-time picture of figure 25 is not the same as a naked singularity such as may form in gravitational collapse—unless it is censored. Like the usual black hole singularity, the big bang singularity is a point in time: a horizontal line in the Penrose diagram. A naked singularity, by contrast, is vertical.

The cosmic censorship conjecture is also important from a mathematical perspective. There are, as mentioned, explicitly known exhibitionistic solutions that show their singularity nakedly. But they are of utmost shyness: In all known cases it has been demonstrated that the minutest disturbance, a slight change in the initial state of the collapse, causes an all-covering horizon to form—a truly effective censorship. In general, however, this behavior of cosmic censorship has not been proven yet even though it was formulated four decades ago, in 1969, by Roger Penrose. Ever since, mathematicians and

25. Penrose diagram for a universe starting with the big bang singularity represented by the dash-dotted line at the bottom edge. Although this singularity is not covered by a horizon either, it is completely different from the naked singularity in figure 24 in that it takes the form of a horizontal rather than vertical line. At the top, the diagram is not drawn completely, since there are different possibilities for the future: As described in chapter 1, depending on the precise form of matter, the universe can either recollapse to a singularity, in which case the diagram would be closed at the top, or expand forever.

physicists alike have tried to prove it. It has initiated many an important mathematical development, and to this day it remains one of the big challenging problems in general relativity. If one considers that this problem is about principal insights into the predictability of everything happening in space-time, one clearly sees in an impressive way the far-reaching importance of general relativity, and of the current research on its open questions.

ANALOG GRAVITY: BLACK HOLES OF THE ORDINARY

If cosmic censorship is taking place, we can analyze black holes only by their horizons. While a direct investigation with probes is most likely beyond reach, general properties of horizons can possibly be studied in the laboratory. The behavior of space-time itself would not be studied in the analysis, but instead matter in some media that can lead to analogous phenomena is used. The first to notice this was Bill Unruh in 1981, who likes to compare a black hole to a powerful waterfall:[4] If the water drops down fast enough, faster than the propagation speed of waves or of sound in water, then an observer—say, a fish communicating by sonar—falling down with the water can no longer transmit signals through the water back to another one staying safely at the upper end. One can often watch this directly as shown in figure 26, for the foaming turbulence down the falls cannot propagate against the stream, upward to where the water remains calm. Rapid waterfalls are only the simplest example for the observation that exotic-seeming phenomena of space-time in general relativity can have analogs in common media such as liquids. Investigating geometric phenomena by means of condensed-matter physics is called *analog gravity.* (Which, it should perhaps be noted, is not a precursor of "digital gravity," as might be supposed.)

While a horizon, in this sense, can easily be manufactured, more subtle, challenging problems related to quantum effects are of experimental interest. For black holes, among these issues is Hawking radiation, to be discussed in detail in the next section. In analog gravity with material media, researchers have the advantage that quantum effects of matter have already been widely analyzed for other reasons. In place of gravitational waves as excitations of space-time, which can reach us from distant sources, in condensed matter systems we have *phonons.* These are collective, mutually coupled oscillations of atoms in a solid or a liquid that, owing to the stiffness of the medium, propagate with a certain speed: A group of swinging atoms can, via binding forces, excite neighboring atoms to swing along.

Since oscillations of a single atom are quantized, as described by quantum mechanics, so are collective oscillations. Their intensity cannot change continuously but only in discrete steps by creating more phonons through atomic vibrations. Here we have an atomic, quantized picture of sound, much as photons give us the quantized picture of light. For gravitational waves one expects a similar quantized view, but would have to employ the difficult quantum gravity for a complete derivation. While their structure remains incompletely known, the name *graviton* for these atomic excitations of space-time already exists. In loop quantum gravity, the final picture might be very close to that of condensed-matter physics; there is indeed a discrete atomic structure of space-time. Its excitations, once they are mathematically understood, should result in gravitons. Initial investigations have been done in quite some detail by Madhavan Varadarajan.

Waves arising from the propagation of an oscillation can be used to transmit signals, provided there are methods for a targeted creation and a precise perception. For light and sound, we use seeing and hearing. On a microscopic level, individual photons of light or phonons in the air are perceived and interpreted in their totality as messages. These exchange particles also have a fundamental meaning in the understanding of physics, for they form the elementary picture of a force. In the case of phonons, this force is the elastic one: A deformation in a solid or an increase of pressure at one place in a liquid propagates as a wave from its starting place and eventually influences distant regions. Deformations and pressure increases occurring there can be traced back, in the macroscopic interpretation disregarding elementary processes, to the action of a force.

Similarly, photons are the elementary constituents of light, which in enormous quantities form the complex signals we often use to communicate. Light propagates even in empty space; no material force is transmitted by photons, only the electromagnetic one. Gravitons have been postulated as the elementary exchange particles of gravity. Despite the incompleteness of its theoretical formulation in

26. A waterfall illustrates a horizon where waves, as sound signals in water, cannot propagate upward against the falls. One can see this by the placid run of the water at the top, remaining undisturbed by the turbulence that builds up at the bottom. Analogously, there is a horizon around a black hole from behind which light cannot escape. (Leura Falls, Blue Mountains, Australia.)

quantum gravity, this notion, in contrast to Newton's efforts, nevertheless shows a clear advantage: Gravitons as messengers must first travel through empty space between two places; the force they transmit cannot act instantaneously. A consistent formulation of elementary gravitons as quanta of gravity would automatically resolve Newton's reservations about his own formula.

A consistent formulation of the force picture, as it is given in its classical description by general relativity, is very complex and often weighed down by apparent paradoxa. Again, the existence of black holes clarifies the problems: A black hole is the source of a strong gravitational force by which it influences masses in its neighborhood. The force is so strong that most galaxies are bound into rotating discs by gigantic black holes in their centers. Only by the gravitational influence on their neighborhood can black holes be detected, for light cannot escape. But if gravity is supposed to be nothing but the force transmitted by the exchange of gravitons, why, then, can gravitons, messengers analogous to the photons of light, leave the black hole to bring notice of the strong central force to the neighborhood? Reconciling the elementary picture of gravitational quanta with the powerful processes bending even space and time is one of the main difficulties of quantum gravity.

Back to analog gravity: As already mentioned in the description of horizons, quantum theory of matter near a black hole horizon leads to a creation process—Hawking radiation. Correspondingly, at a horizon in a medium, one should expect a similar creation process of phonons. Since one can easily build an analog horizon, and since phonons are nothing but detectable sound, one can test the general process in this way. Ralf Schützhold, for instance, has recently proposed experiments. The field is currently a vibrant one, with major contributions also from Carlos Barceló, Stefano Liberati, Matt Visser, and Silke Weinfurtner.

Although it may seem easy, one cannot simply hear the sound of analog Hawking radiation, for it is not intense enough. The medium must be isolated from other external sources, a challenge complicated by the production of a horizon whose existence requires high velocities in some parts of the medium. Hopes rest on exotic forms of matter, called *Bose-Einstein condensates*, that in some materials arise at very low temperatures. They have the advantage of a very small sound speed; tiny velocities of the medium are then sufficient to confine sound waves to a given region. Moreover, this kind of matter exists at low temperatures, easily lower than that of the phononic Hawking radiation. In contrast to black holes in the universe, Hawking radiation is bright (or, rather, loud) in comparison with the background radiation caused by the ambient temperature and thus should be more easily detectable.

These experiments are only in the planning stage. But they definitely provide the possibility to probe properties of horizons in a secure laboratory before undertaking risky tests of black holes in space. No dangers arise, for there is no singularity and the horizon automatically disappears when the liquid comes to rest. If analog Hawking radiation can be confirmed and precisely measured in this way, it will increase confidence in mathematical methods used to compute it. In addition, there are many foundational problems in the case of Hawking radiation that are not yet fully understood. Instead of using mathematics, which has been attempted for several decades with slow progress, successful measurements could allow us to directly ask Nature herself.

QUANTUM THEORY OF BLACK HOLES:
MAKING THE MUTE SPEAK

Black holes provide a treasure trove of questions for which the quantum theory of matter or gravity is of crucial importance. This is the case for the region of the horizon as well as for the final singularity of the collapse. In particular, we must pose the question of what a black hole really is, for general relativity, due to the existence of a singularity, does not give us a complete picture.

COSMIC EVOLUTION? BIRTHING WORLDS

It is clear that black holes exist in the universe, making it all the more shocking that we do not fully know them. The question is not only what kind of astrophysical object we have here, just as the nature of gamma ray bursts is not completely clarified. Black holes pose an entirely different problem: It remains to be determined whether they are just dense and highly curved regions in our universe, like many other heavy objects, or splitting-off processes of space-time branching out into a daughter universe. The decisive question is then: Does the horizon mark a transition region into a different world accessible only through it, or is it merely a transient stage for a region within our universe?

These two possibilities can be illustrated once more by their Penrose diagrams if one graphically extends those beyond the horizontal line of the singularity. Within general relativity, this is not possible, for the singularity is a borderline, but it is useful to know all available options for subsequently analyzing them in quantum gravity. As illustrated in figure 27, this region could be connected to the old black hole exterior, and the black hole would just represent an extremely dense core. Alternatively, it could be disconnected, in which case—as the possibly more spectacular alternative—the whole new, unexpected world of a daughter universe would open up behind the singularity. The black hole would be a portal marking the fork between mother and daughter universe, and it would have a horizon cutting the umbilical cord.

Precisely how to extend the classical picture can be determined only by means of a theory of quantum gravity, for this governs (or so one expects) all events at the singularity of general relativity. A final clarification of the question remains open, but one can already use some of the known properties of quantum theories of gravity to speculate; and the most exciting speculation is, of course, the more spectacular possibility mentioned above: the creation of a daughter universe.

Theorists have repeatedly aimed their thoughts at this issue, but Lee Smolin, one of the founders of loop quantum gravity, has approached it with particular creativity. He assumes that this daughter universe indeed grows up to be a whole big universe like ours, and even that our universe was born in this way from an ancestor. Such birthing events are to be distinguished from a bounce as in cosmology, a simple turnaround in the expansion of a single universe. Through black holes, one universe would at some point split off into two. And if this is repeated, one obtains a strongly branched space-time of decidedly complex structure. Every black hole in the cosmos

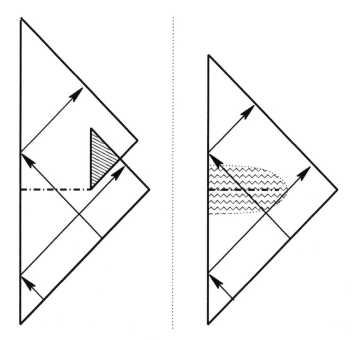

27. Alternatives of a black hole with prevented singularity. Left: Behind the singularity there is a daughter universe without contact with the previous black hole exterior. (The hashed region marks an overlap in the visualization where, however, no contact between the space-times exists.) Right: The classical singularity is merely surrounded by a high-curvature region (drawn wavily), penetrated in quantum gravity. Here, the diagram no longer differs significantly from that of a common compact object as in figure 22.

can give birth to its own daughter universe, and each daughter universe can, through the black holes it contains, produce granddaughter universes, and so on.

Another difference between universe birthing and the cosmic bounce is the fact that a discussion of daughter universes necessitates a theoretical viewpoint outside space-time itself. As internal observers, which we really all are, one can choose only between remaining with the mother universe, outside all black holes, or devotedly throwing oneself into one favorite daughter universe. One can never oversee the entire space-time as an observable and physical object; instead, when toying with several separate daughter universes, one would take a metaphysical standpoint outside any observability whatsoever, even in principle. Already in cosmology we saw immense difficulties in watching the universe before the big bang, although in that case it was possible at least in principle. But viewing several daughter universes at the same time is fundamentally impossible.

Here, Smolin's important insight enters the game. He argues that there could nonetheless be observable consequences of such repeated branching-off processes. One merely has to make one more assumption, namely, that during each splitoff, parameters of the universe such as constants of nature change slightly. Concrete reasons do not exist, but this is up to quantum gravity to decide; we shall thus accept the assumption for now. (In string theory, many "constants of nature" are established as values taken by fields that may vary over long durations, providing a possible scenario for such changes to be combined with Smolin's idea.)

Smolin goes on to point out that one is dealing here with an evolutionary process akin to that in biology. According to Charles Darwin, the immense multitude of species of life on earth emerged through evolution founded, in simple terms, on the two principles of mutation and selection. *Mutation* means that offspring differ slightly from their parents; *selection* then unflinchingly chooses those individuals with the best properties for a given niche and strikes out others. This happens not haphazardly, but by an apparently rational process: In this sense, the most successful properties are those that promise the individual the highest amount of offspring making it into adulthood, for this ensures the survival of the individual's genetic "self" beyond its own death.

Space-time could be ruled by similar principles. If properties of each new daughter universe change slightly, one has mutation. Selection occurs, too, because the reproduction rate via black holes in a daughter universe is subject to the physical laws that are valid there. (In this context, selection is understood in a statistical rather than existential sense, since individual universes cannot directly compete for resources. Rather, they compete for the statistical dominance of their properties in the multitude of all daughter universes.) For instance, the Chandrasekhar limit, the maximal mass of a white dwarf, depends on constants of nature such as Planck's constant of quantum theory or Newton's constant of gravity. When this limit is lowered by changing the constants, black holes could form at even smaller masses. Daughter universes with suitable parameter values— values that lead more easily to black hole formation—will produce more granddaughter universes than others and dominate as matrons in the metaphysical cosmos. With these lines, a picture of cosmic evolution is drawn that may account for the astrodiversity of objects in our universe—not only directly for black holes but also for related phenomena such as galaxies or possibly gamma ray bursts.

Smolin's proposal has a notable feature: It is testable, at least in a statistical manner, and thus it is of scientific quality. In this regard it differs from other versions of so-called *multiverses,* patchworks of several space-time regions separated from one another and mutually inaccessible. Statistically, we should, after all, find ourselves in a typical part of the multiverse, in a daughter universe of a kind appearing especially often. Analogously to the biological example, the fittest individual universes, those with the most offspring, are most common. In this scenario, offspring come through black holes, the branching points of the multiverse into new daughter universes. If our universe is a typical one, it must produce many black holes, something it apparently does: Most galaxies carry an extremely massive one in their center and many smaller ones outside the center. One can get a more quantitative handle on this picture by estimating the characteristics that constants of nature must possess in order to produce black holes in large numbers. If, for instance, Newton's constant were too small and gravity too weak, then the clumping of matter in the early universe would proceed too slowly, leaving just a widely spread-out gas left in the universe. Or neutron stars could perhaps remain stable for all masses, not allowing the further gravitational collapse into black holes.

In principle one can see whether our universe is indeed a contender in the interuniversal fertility contest, if the number of black holes is the measure of success. Calculations and statistical estimates, however, are, to put it mildly, difficult. The few derivations done so far remain disputed. Regardless, Smolin's conjecture is based on the assumption that the singularity of a black hole in quantum gravity is indeed eliminated by the universe's branching off into a daughter universe. This assertion is as hotly debated as it is difficult to test, but after the following section it will bring us back to the topic of black hole quantum physics.

HAWKING RADIATION AND INFORMATION LOSS: ALL UP IN SMOKE?

On the way to a quantum theory of black holes we first need more details about Hawking radiation. Its derivation does not rely on any version of quantum gravity; rather, it uses quantum effects of matter in a universe that is curved by space-time transformations but is not quantum itself. Even here, as during inflation, the special, trembling form of space-time causes particles to be created from the vacuum. As seen in figure 23, the diagram of a black hole has a form entirely different from that of other astrophysical objects in figure 22 or of empty space in figure 21. Differences are noticeable especially at the horizon and beyond, in the black hole interior; near the horizon we have a very different structure of space-time than at safe distance from it.

As in the analogy of the seabed's influencing the propagation of waves in quantum theory, or in the example of inflationary generation of matter from the vacuum in cosmology, the form of space-time is responsible for the behavior of a quantum mechanical wave function of matter. Wave functions of matter are influenced by the unsteady ground of space-time near the horizon such that partial waves with particle character split off. In the particle picture, this happens through the creation of particle-antiparticle pairs, one partner falling into the black hole, the other one escaping as part of Hawking radiation.

With the escaping particles, the black hole loses energy and mass. Its horizon, with a radius proportional to the mass, shrinks. Initially, all this happens extremely slowly because the temperature of a heavy black hole is very low and not much energy is emitted. A large black hole has a horizon far away from the singularity; near the horizon, deviations of space-time from that in empty space are realized in a much smaller fashion than for lighter black holes. In the seabed picture, the ground near the large horizon of a heavy black hole is more similar to empty space than that around the small horizon of a lighter black hole; fewer waves, or particles, are being created.

Unless more energy from the outside happens to fall in than is radiated away by the Hawking effect, the horizon radius must continue to shrink. For black holes in the usual mass range, such as can be found in the universe, even the weak cosmic background radiation right now provides enough energy to stabilize black holes against Hawking evaporation. With further expansion of the universe, however, the background radiation will cool down more and become ever weaker. Lighter black holes will start to evaporate sometime in the distant future. Only engulfing neighboring matter such as stars or intergalactic gas can save them from evaporation, but this food will eventually be used up. Viewed over the whole history of the universe, the case of evaporating black holes is thus rather realistic.

When it starts to evaporate, a black hole shrinks and heats up more and more. This enhances the evaporation process and must, if it is simply extrapolated, lead to the disappearance of the horizon after some finite time. Such a theoretical extrapolation, however, is not valid, because the horizon, once small, eventually comes close to the singularity in the strong curvature region. At this stage one needs a quantum theory of gravity, which Hawking did not take into account. His calculations were instead based on general relativity describing the black hole and its horizon, and on the quantum theory of matter in this space-time—but not on that of space-time itself.

On its own, the Hawking process cannot help in the decision as to what the black hole singularity really means. The singularity does not simply disappear into thin air (or faint radiation); the quantum theory of matter alone cannot provide a solution to the singularity problem. Instead, the evaporation reveals further problems with the classical picture, which are useful in stimulating a later resolution in quantum gravity by providing different viewpoints.

Most widely discussed among these problems is the so-called *information loss paradox.* All possible things can fall into a black hole and take with them an immense amount of information. What comes out after the Hawking process, however, is merely black body radiation from the neighborhood of the horizon, which knows nothing about the contents of the black hole. Such radiation is among the most boring, information-poor things in physics.[5] A single parameter suffices: the temperature. If it is known, the whole intensity profile of the radiation is already determined; it must then be of the form derived by Max Planck when he unknowingly kick-started the events leading up to quantum mechanics. Most of the radiation is emitted within a frequency range determined entirely by the temperature.

Black holes would constitute gigantic information shredders, representing a big problem in quantum theory. Although one does not know the precise equations describing the complete quantum state or the wave function of a black hole, one can show that no equations as they otherwise appear in quantum theories can lead to such an extreme destruction of information. Here it is important to emphasize that the information is not just difficult to access; it cannot be reconstructed even in principle. If a book is burned, the information it contains is no longer readable, but with subtle techniques it may still be recovered from the charred pages. Information destruction in a black hole is much more thorough: There are simply no information carriers by which lost data could be saved.

Perhaps the problem is more tangible when formulated by means of conserved quantities. Certain numbers in physics, characterizing the total matter content of the universe, are not allowed to change in any known process. An example is energy, which is also constant in the Hawking process. Other conserved quantities are electric charge, or the total number of protons, neutrons, and heavier partners of those particles, called the *baryon number*. Their antiparticles count negatively, keeping the whole number unchanged during pair creation, for instance of a proton and an antiproton. Now, nothing prevents us from sending more protons than antiprotons down into a black hole; this is likely to happen anyway, since as far as we know, there is much more matter than antimatter in our universe. A black hole would then contain a surplus of protons. Hawking radiation, on the other hand, treats matter and antimatter democratically, holding that it is just as likely for the particle half of a particle-antiparticle pair near the horizon to fall into the black hole and the antiparticle to escape as radiation as vice versa. (Electric charge might distinguish the partners, by preferred attraction or repulsion, but black holes are not strongly charged.) Should the black hole evaporate completely and leave only Hawking particles, one important conservation law would be violated.

One can solve this problem only by assuming a stable remainder of the black hole at the end of evaporation. But what object could this remainder be, and why should evaporation then stop? Only quantum gravity phenomena can help, for general relativity allows no end to evaporation and no remainder either. Once matter has collapsed, the classical theory admits only the possibility of black holes with a Hawking-evaporating horizon. A compact core left after evaporation could be explained only by a quantum theory of gravity. But if quantum gravity, as often expected, becomes relevant only at the scale of the Planck length, this residue must be tiny. The question then arises whether it can, even if it is stable, shelter a sufficiently large part of the fallen-in information to make the process consistent with quantum theory.

As with the big bang singularity, we cannot evade quantum gravity and an investigation of its details. Again, we are dealing with a problem of time: Space-time has a border that, behind the horizon, does not grant enough time for, say, light to escape. Discrete time can eliminate the big bang singularity, providing additional time beyond it; and with black holes, this can perhaps help too. After all, the point-in-time singularity of a black hole appears quite similar to that at the time of the big bang—except that a black hole takes the space only of a limited region, not all of space. Should one obtain more time, a new region of space-time may open up: a new world—either a daughter universe or a new part of our universe after the black hole's evaporation—of which general relativity knows nothing whatsoever.

QUANTUM GRAVITY OF BLACK HOLES: LIFE MUST GO ON

If black holes emit energy, they can be considered to be analogs of the excited atoms that tell us much about the discrete energies in quantum mechanics. Atoms in an energetic state above the ground state can emit surplus energy to transit to the ground state. Possible amounts of radiation energy, measured in the atom's emission spectrum, provide insights into atomic structures in quantum mechanics. Black holes, similarly, are gravitationally bound systems whose radiation can give hints about their structure. They emit energy by the Hawking process, as yet unmeasured but computed. In studying Hawking radiation, one makes use of general relativity and the quantum theory of matter, and one obtains a result not fully consistent within those theories as a result of the information loss problem. In particular, the end state after Hawking evaporation cannot be understood in this way. The final state indeed looks analogous to the atomic ground state, to be described only by means of a complete quantum mechanics. To understand the finale of Hawking radiation, a quantum theory of gravitation, space, and time is needed—not one of matter alone as in Hawking's calculations.

String theory has difficulties with the dense regions around the singularity, because of the lack of a detailed understanding of space-time structure, and has not provided a concrete picture for what black holes might look like in quantum gravity. Rather, it tackles such questions by looking at the singularity from a safe distance, in fact from infinitely far away: It provides so-called holographic descriptions in which everything that is going on somewhere in space-time is represented by structures at the edge of space-time out at infinity. (Similarly, a hologram presents a three-dimensional scene on a two-dimensional screen.) Mathematically, this process is elegant, but it is also indirect and not very concrete. With these constructions, string theorists have vaguely concluded that no information is lost in a black hole since the screen infinitely far away does not lose information. But the question remains whether the description by the screen is complete; if it is not, information in the full state may still be lost.

String theory models have been more successful for a description of the horizon than of the singularity of a black hole. Although the horizon is not a material surface, it should have a microscopic structure owing to the atomic nature of space and time. One can then try to count the number of building blocks, or the number of all possible ways to construct a black hole of a certain size. As undertaken by Andrew Strominger and Cumrun Vafa, one can count all possible states (the so-called *black hole entropy*) in string theory. Such calculations are often seen as the first test that any quantum theory of gravity must pass, and they have been performed successfully in string theory at least for certain types of highly charged black holes. These objects are not quite realistic, for a charged matter distribution under collapse would eject charged particles and be almost neutral once a black hole forms; nevertheless, the successful calculations in string theory present an important consistency result. This encouraging outcome is, however, no proof that the theory is correct, and it does not necessarily distinguish it from others. Loop quantum gravity, too, shortly after string theory, succeeded, by a rather different calculation, in deriving and analyzing black hole entropy—and that for a much more realistic class of uncharged black holes.[6]

More interesting, but also more complicated, than the number of states for a fixed-size black hole is its behavior when it exchanges energy with its surroundings—either by absorbing more matter or by emitting Hawking radiation. In the discrete picture of loop quantum gravity, black holes present a kind of atomic system built not from blocks of matter but from chunks of quantized space-time. A Hawking photon is emitted when the black hole changes its state of energy or mass. Hawking radiation is described by a theory of waves in a given space-time, similar to Planck's formula for the radiation in a black box. To understand the emergence of this radiation better, one must add a theory of the walls of the box and their interaction with the radiation by emission and absorption of photons. For the box, Einstein had achieved a successful theory of emission and absorption by his explanation of Planck's formula.

For a black hole, one needs a quantum theory of space-time near the horizon because that is where the Hawking particles are generated. A final explanation must use quantum gravity. Once the explanation is available, one can perform spectroscopy of black holes analogous to that of atoms and molecules, which was so important when quantum mechanics was developed. What will remain out of reach for a long time, however, is a comparison with observations, for they would have to measure not just Hawking radiation itself but also its details such as the intensity distribution over different wavelengths. For astrophysical black holes, our universe, despite all its expansion and dilution, is simply still too warm, dwarfing Hawking radiation by the cosmic microwave background.

Still, even theoretical spectroscopy allows many interesting questions. The atomic structure of space and time, as it arises in loop quantum gravity, indicates that the surface area and the mass of black holes, related to each other by the Schwarzschild radius, can take only discrete values. Based on general principles, this was postulated by Jacob Bekenstein shortly before Hawking's calculations. Loop quantum gravity by now presents a concrete proposal for the mathematical form of the discrete sizes constructible by stacking together atoms of space. But one still has to see which transitions happen when Hawking radiation arises, or how atoms of space are transformed into radiation particles. Of particular interest is the question of whether a black hole has a ground state akin to that of an atom, and what space-time could represent this state. At the end of the Hawking process one should, by the atom analogy, expect the black hole to be in the state of lowest energy; its form is thus decisive for the question of whether Hawking evaporation indeed destroys all information, or whether there is a compact core for storage.

Then again, the analogy to atoms may in the end lead to wrong conclusions even though it can describe the first steps of Hawking evaporation. For as the black hole becomes smaller and smaller, the

Hawking emitting horizon approaches the singularity and its high-curvature region ever more closely. Nothing comparable happens in material atoms, where instead energy simply decreases when an atom approaches the ground state. In the ground state, properties of quantum theory are very important, for otherwise the classical stability problem could not have been solved; but no phenomenon of high energy density, as the singularity presents it, arises. To answer the crucial questions about black holes, the singularity and its fate in quantum gravity must be understood.

As in cosmology, there are two components of loop quantum gravity that can make the singularity disappear: repulsive forces, and the availability of more time by time's discretization. With black holes, we cannot rely on such symmetrical situations as in cosmology, because homogeneity would be too strong: We require a distinct point as the center of rotational symmetry, and the gravitational force changes with the distance from that point. Calculations are more complex, and the possibilities offered by quantum gravity are not yet reliably investigated. But here, too, there are not merely clear indications pointing to the existence of repulsive forces and of additional time; even concrete models without singularities exist. For Abhay Ashtekar and me, this was a welcome opportunity in the early years of loop quantum cosmology, allowing us to probe the framework's consequences in more general contexts.[7] Even though we still do not know with certainty whether the singularity is replaced by a daughter universe or whether black hole remainders stay in the existing universe, possible processes can nevertheless be clearly delineated.

The fundamental phenomenon is once again the discreteness of time. Not all possible points in time are allowed; instead, time jumps in fixed, precisely controlled steps. Although the step size is very small—playing hardly any role outside the black hole, at the horizon, or even in most of the interior of large black holes—it does become essential near the singularity. In contrast to the continuous time of general relativity, which is doomed to stop after a finite interval and to break down with the theory itself, discrete time keeps on running. The elimination of some points in time on microscopically small scales implies the occurrence of a new, much wider, macroscopic time frame beyond the classical singularity of a black hole. In a Penrose diagram, this is represented on the left-hand side of figure 28.

For the extension of time, it is crucial that the singularity inside a black hole is a point in time but not, like the center of a star, a point in space. Only this way can it be possible for discrete time to jump across the singularity and unlock a new world in the future.

We have here one of the important realizations of consistency, guiding theoretical developments in the absence of direct observations: Precise details of the classical theory play a decisive role for the concrete mechanism of the quantum theory, extending it in a way that could not have been foreseen while quantum gravity was set up. Black holes in general relativity, as they arise from the collapse of matter, are distinguished by a point-in-time singularity behind the horizon, penetrated by the discrete time just like the big bang singularity of cosmology. If the singularity of a black hole were a point in space like the center of a star, time would run parallel to it and have

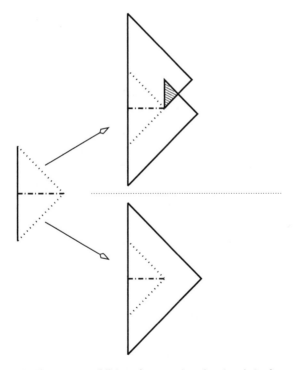

28. The two possibilities of connecting the singularity-free black hole interior of quantum gravity with the exterior. Either there is a branching point to a daughter universe beyond the dash-dotted singularity, or the interior simply sits in a space-time that surrounds it completely.

no chance of piercing and eliminating the singularity. Naked singularities would persist in quantum gravity—unless they were already forbidden by cosmic censorship in general relativity.

Classical characteristics and those of quantum theory play important, closely entwined roles, illustrating the elegance and consistency of the ideas involved. Quantum gravity is effective and unforgiving in the elimination of dangerous singularities, but at the same time economical when it comes to singularities already dissolved classically. The fate of singularities in quantum gravity, despite that theory's current incompleteness, provides a test to be taken seriously, one that has been passed in the form of early mathematical derivations. The more such passed tests pile up over time, the stronger one's trust in the theory becomes. Should a test fail, the first recourse would be to improve the theory in such a way that this test can yield a positive result while the earlier test results remain valid as well. Should this not be possible or should the modification make the theory look too artificial, one would eventually drop it altogether. Even without observations, there is thus strong selective pressure on theoretical speculations.

If the singularity as a point in time is now eliminated, there is a time with space behind the singularity or, seen from outside the black hole, a universe after Hawking evaporation of a black hole. A crucial question is whether the new space-time after the singularity in the interior is connected with the space-time existing outside—or, as asked above: Does space-time branch off into a daughter universe inside a black hole, or will the collapsed matter be thrust back to the exterior? This question is important for the physical picture and for potential observations well into the distant future; indeed, our whole view of the cosmos and its theoretical underpinnings are put to a tough test by this question.

In contrast to traversal of the high-curvature region around the classical singularity, a precise analysis of all details of temporal changes as they play out among the elementary constituents of discrete space-time is necessary at the would-be branching point. During a branching off to a daughter universe, space-time must, after all, separate between two neighboring space-atoms, connected initially but no longer at the following time step—after the singularity. At the present stage, a complete answer is difficult owing to the incompleteness and mathematical complexity of the theory. Yet there are indications of what the answer should look like once a complete and consistent theory becomes available: Namely, there is no branching off into a daughter universe; instead, the interior will be united with the exterior after the horizon evaporates.

Then the collapsed matter will reappear, rent asunder—possibly in an explosion widely visible from afar in the universe. The region of extremely high energy density and curvature around the classical singularity suddenly becomes accessible from outside, which is bound to represent a powerful cosmic event. Information loss such as Hawking initially feared does not occur, since with matter the stored information would also return. It may be highly deformed and difficult to decipher, for it will have had to traverse the inner region of high curvature, but in principle it is still there. And this is all that is required for the consistency of the theoretical framework.

Preliminary demonstrations of this behavior rely on characteristic properties of atomic space-time dynamics as it can be seen in loop quantum gravity. There might indeed be a rending of neighboring space-atoms where curvature is very high and quantum aspects dominate; the predecessor space-time could split into two separate regions. Depending on the exact form of the dynamics, which is yet to be found, there are two possibilities: The briefly separated parts directly reconnect in the next discrete time step, or they remain divorced forever.

Only in the second case can a branching off into a daughter universe occur. But the problem is that the point-in-time singularity of the black hole is spatially extended, just as the usual center of a star is not spatially but temporally extended. For a splitting off of the black hole interior into a daughter universe, it would be necessary for space-time to tear only at one end of the classical singularity—at one side of the high curvature region in general relativity, the right endpoint in our figures—but to remain intact elsewhere. Otherwise, the interior would indeed separate at the horizon and immediately disintegrate into numerous completely fragmented space-atoms: a cosmic miscarriage, in no way producing a daughter universe.

Since the high-curvature region is subject to the laws of quantum gravity not just at its edge but over its whole extension, a tearing at the edge comes just as easily as a ripping off anywhere in the interior. There is, after all, extremely high curvature everywhere, together with strongly quantum behavior; no reason exists for a special treatment at the edge. No longer does the theory have to decide between a branching off to a daughter universe and an opening up of the black hole back into the original space-time, but instead, and more dangerously, it must decide between the latter case and the interior's complete disintegration.

How can one rule out a disastrous disintegration? Fortunately, this can be done relatively easily: Quite by luck, the interior of a nonrotating black hole obeys the same laws as are found in cosmology, based on the assumption of exact homogeneity. A black hole's whole space-time, especially its exterior, is not at all homogeneous, because the gravitational force strongly varies with the distance to the central singularity; but inside the horizon, this radial dependence turns into time dependence. Again, the point-in-time nature of the singularity is crucial: Approaching the horizon, we decrease the radius, a spatial coordinate. But once we pass the horizon and go on toward the singularity, we move to the future where the singularity lies, ahead of us in time. Thus time, but not a spatial coordinate, changes when we move toward higher curvature. And it indeed follows from the mathematical properties of Schwarzschild's solution that the interior can be considered homogeneous but—as in cosmology—time-dependent.

Homogeneity significantly simplifies investigating the dynamics, since all places in space can be considered at once. Just one equation, as on page 118, is to be solved for the whole interior, instead of uncounted numbers for all the spatial atoms. We already know the result from quantum cosmology: The singularity of the classical theory is traversed without the disintegration of space. For a black hole, this means that its interior must open up to the exterior after the horizon evaporates, and thus will become visible for an observer.

Incomplete knowledge of quantum gravity notwithstanding, there is a unique prediction. One must keep in mind, though, that we had to use different assumptions and indications in this chain of proofs, still to be buttressed by detailed calculations or even the redelivery of missing theoretical elements. The most important step will be to solder the interior dynamics, for which we used cosmological results, to the so far poorly understood exterior dynamics. We started our argumentation with the required completeness of an encompassing theory, according to which the homogeneous interior dynamics can consistently be combined with the exterior dynamics describing the horizon's evaporation. The picture as described may thus be taken as a prediction of current ingredients of the theory, whose consistency can be further substantiated once extended calculations become available. If this process is successful, the result should also be generalizable to rotating black holes, where our present arguments lose their validity as a result of the inhomogeneous interior in this case.

The lucky relation of the nonrotating black hole interior to cosmological equations also allows the development of a preliminary but concrete picture of the black hole and its fate after evaporation. Again, the existence of a repulsive force is crucial, as it arises at high compressions of the collapsed matter when quantum gravity is considered. Matter, then, cannot collapse completely to the point in time of the classical singularity; it is instead being dispersed out after sufficiently strong compression. As in cosmology, the limited storage space of discrete time is responsible for the emergence of this counterforce. Maximal compression is obtained in the black hole's center, allowing the singularity in general relativity but not in quantum gravity. Seen from outside, this point in time is marked by the horizon's disappearance by evaporation. As illustrated in figure 29, the

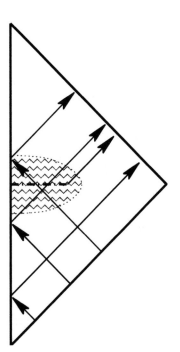

29. Representation of astronomical observations at a nonsingular black hole. Moving in time along the top right border, we first encounter radiation that had passed the black hole before it formed (bottom arrow). As the first signal from the black hole, one sees Hawking radiation escaping from the edge of the classical singularity. Afterward, all radiation from the black hole region has passed the high-curvature stage. This radiation carries remainders of matter that has fallen into the black hole and could contain immense amounts of energy. Such a scenario has been obtained in loop quantum gravity and also recently, based on different ingredients, by Gary Horowitz using string theory. Independently of quantum gravity, a picture of this form had been anticipated by Peter Bergmann, Tom Roman, and Sean Hayward.

interior, in contrast to the classical scenario, will again be visible from the outside in the form of hot, extremely dense matter, driven apart by the repulsive forces of quantum gravity as in a mini-bang. From the outside, one would first see the rest of the Hawking radiation, soon overpowered by the exploding interior's glare.

As already mentioned, this picture remains to be supported by further calculations—required, for instance, for a quantitative estimate of the energies involved. Even though it is quite certain that the unfolding interior would radiate much more energy than the final stage of Hawking radiation, it could be just a small fraction of the whole of collapsed matter. It takes, after all, a long time for heavy black holes to evaporate. In this time, a large fraction of the total energy can already be emitted by the Hawking process, even though in brief intervals the radiation is only of low intensity. Before all these ingredients have been combined, the strength and brightness of a possible black hole explosion cannot yet be reliably estimated. But given the extremely high curvatures in the interior, far surpassing anything in the universe except the big bang itself, one must expect a gigantic event. Unfortunately (or perhaps luckily), we will have to wait a long time for this to happen in our universe. For now, black holes are protected from evaporation by cosmic background radiation, despite its weakness.

At first, the gravonomist on duty did not consider the speck of light in the black night sky exciting, but it was a welcome distraction in his otherwise dull routine. Eruptions of this kind had long lost their initially spectacular flair, even though they represented the sole remaining cosmic events visible to the naked eye. Everything else was long extinguished, after the expansion of the universe had made other galaxies and then even neighboring stars drift too far away. The unrelenting expansion had, however, also made the cosmic background radiation cool down more strongly, and at last facilitated the evaporation even of massive black holes. Long after collapse and compression, their matter now reappeared in immense explosions. They were the main sources of gravitational waves, whose detection by gravonomy had replaced the traditional mapping of the sky by astronomy.

Scientific investigations of the universe, however, became ever rarer, and gravonomy was tightly held by the military. The origin and fate of the universe, growing larger and emptier, appeared clear;

thus scientific interest in gravitational waves and other, weak cosmic messages diminished. Instead, the old observatories served for military surveillance. After the depletion of traditional energy sources, the extraction of energy from black holes by the Penrose process became the last elixir of life. To that end, some of the naturally existing black holes had been artificially enlarged by a process which came to be called "horizon broadening." Initially, as a by-product, humankind had used the opportunity to eliminate unpopular civilizations unfortunate enough to have settled near a black hole, but this practice was soon largely banned. Among all civilizations of the universe, as always, humanity had distinguished itself by its destructive power. It also had had the luck of having early in its history developed a successful theoretical description of space-time by general relativity. (By humans' own fault, geo- and science-political circumstances had for several decades hampered their gravitational research, but these played no role in the following centuries.) This knowledge and its application was used by humankind to consolidate its dominance, after which they officially banned horizon broadening. Only a few rogue galaxies were still suspected of using horizon broadening, and they were monitored to detect the gravitational waves inevitably emitted during a broadening.

Energy companies had at first supported wide-scale horizon broadening, too, but for another reason: They had realized the danger of the continued expansion of the universe. The consequent dilution of matter would, very slowly but unavoidably, not allow even heavy black holes, the last sources of energy, to exist for long. They lobbied for the Buchert decelerator: Dark energy as the source of accelerated expansion had long been recognized as a consequence of inhomogeneity in a universe having become more structured after the big bang. In the belief that the inhomogeneity would weaken when numerous galaxies disappeared in black holes, horizon broadenings were initially undertaken without restriction.

This was doomed to be an immense failure, for matter fallen into black holes contributes to their mass and does not sufficiently weaken the gravitational force and its inhomogeneous distribution. Except for the expurgation of galaxies, excessive horizon broadenings had had hardly any consequences. Still, they did contribute somewhat to a weakened acceleration, which quite naturally occurred as a result of the dilution of inhomogeneity in an expanding universe. Of course, *the intergalactic government celebrated this as a success of military- and energy-political adventures.*

While such memories from the history books passed through his mind, the gravonomist concluded that his latest observation was a natural explosion after the evaporation of a black hole, and he almost returned to his reading. But this signal was slightly different: When his glance once more glided over the screen, which showed the gravitational wave spectrum in all its details, he noticed something remarkable.

The evaporation and explosion of black holes had successfully been described by stoop theory, established after millennia-long efforts as the theory of quantum gravity. It combined the advantages of all its predecessor theories. Developed just in time, the theory had been tested on the last breaths of the ever more rapidly paling cosmic back- ground radiation: it was a spectacular success. Later on, when the first black holes had exploded, these details too were very precisely described by stoop theory, which was ever since considered the final and fundamental theory of the universe. In the latest event the gravonomist easily recognized the characteristic features of the grav- itational wave spectrum of an exploding black hole. But there was a weak, unaccounted-for high-frequency contribution rising just above the measurement uncertainties. Might stoop theory not be the last word of theoretical physics after all?

The news of possible deviations between theory and observa- tions — the first after many centuries — spread like a brushfire. New calculations were undertaken and modifications of stoop theory pos- tulated, which all turned out to be unsuccessful. Compared to this, the final explanation of the deviations initially looked more down-to- earth: It was a historian who in old records found references to the capsule Kruskal and its long-ago mission flown into the central black hole of an average galaxy. Details were no longer to be found, but now scientists easily managed to trace back the high-frequency part of the gravitational wave measurement to the capsule's signal on its way into the center, to its own demise. And thus the new observation did, after all, contribute to a triumph of the theory, for the first clearly identifiable signal was seen that, unlike light, was able to penetrate the extremely hot and dense central region of a black hole. Once the origin of the signal was known, the observation could clearly be rec- onciled with theory . . .

7. THE ARROW OF TIME
AN AGING WORLD

We fought with the Megadae who are born old,
and grow younger and younger every year, and die
when they are little children . . .

—OSCAR WILDE, *The Fisherman and His Soul*

Time is often annoying. Everything would be so much easier if time would only behave like space. One could then simply take a step back in time to correct a past mistake, or revisit a missed opportunity. Or one could buy some extra time like real estate when the deadlines come too thick and fast. Life is held tightly in time's grasp: At first, everything happens so rapidly and hectically that we take no notice when we pass our own future. For a long while this goes on, but time remains unfazed. At last, what always pushed us on incessantly leaves us behind forever, moving on with those who remain.

Why is time so different from space when both together, according to general relativity, constitute space-time and are even partially transformable into each other?

CONCEPTS OF TIME:
MOTION IN RELATION

Does time not have time?

—FRIEDRICH NIETZSCHE, *The Gay Science*

Time in everyday life bears the deep imprint of its unstoppable (and subjectively nonuniform) progress, clearly distinguishing the past from the future. Crucial here is memory, which makes us recall (and often forget) the past, only in this way showing us the existence of the future: The recent part of the past in our experience is reconstructed as the partial future of our older past. Some events, even unimportant ones, have direct consequences; in the memorized reconstruction, they are given the status of causal relationships. One then expects similar consequences of present events in the future, even if they have not yet happened.

Though common, this concept of time is extremely complex and must be distinguished from mathematical time, which is supposed to describe nature independently of a memory. Time here is like the labeling of a movie's frames, not a player in the action but a mere mark of the physical happenings in their causal order. Shown are either the positions of bodies in classical physics, or the form of the wave function in quantum theory. Even in general relativity this is the case, although time assumes a more active role and can, for instance, come to an end. Time no longer provides absolute marks due to its transformability with space, but marks depending on an observer; it does, however, still determine which events can make others happen and those where no causal connection is possible.

The mathematical picture of time is useful for concrete investigations, but it is idealized. For such a time is not being observed; what we see are rather changes in matter configurations such as the position of the sun relative to the earth, or the hand of a clock relative to the dial. Clocks of any kind are made from matter, subject to natural laws as well as to influence by other matter. One says that a clock keeps good time if the influence from other objects is small, but weak influence always exists; this is the reason why the picture of time completely independent of matter is idealized. Real physical motion is always relational, just as the swing of a pendulum, for instance, is measured relative to the position of hands on a clock. Only matter is ever observed, never time itself.[1] This is especially important in cosmology and its quantum version, as we will soon see. While current conditions on Earth easily allow the construction of useful clocks, there may be regions, in particular in early phases of the universe, where strong interactions between all matter components, including clocks, can never be avoided, for instance owing to high densities. In such cases, one could at best use cosmological quantities themselves, such as the changing volume of the universe or its density, as a measure for time.

Independently of whether in mathematical or physical time, natural laws determine general behavior, while initial values govern special situations. This duality has a direct imprint on mathematical description by differential equations. If, for instance, one provides the positions and velocities of the planets at a fixed initial time, the laws of general relativity determine their whole orbits around the sun and their traverse in the course of time. No difference between past and future exists here: Stopping the motion at some later time, reversing the velocities of the planets, and again using general relativity, one obtains exactly the same trajectories, just in the opposite direction of time. How can memory arise in such a world, distinguishing between past and future in spite of everything?

ENTROPY:
DISORDERLY CONDUCT

Can't bring back time. Like holding water in your hand.
Would you go back to then? Just beginning then. Would
you?

—JAMES JOYCE, *Ulysses*

Most processes of life do not involve just a few clearly defined bodies such as the planets, but rely on the motion of innumerable atoms, impossible to follow individually. To organize such a mess, measurements as well as theoretical descriptions focus on collective quantities of all microscopic constituents, such as the combined volume they occupy, their pressure (as a measure of the total force acting via atomic impacts on a surrounding wall), or their temperature (as a measure of the average velocity of all particles).

Many of these collective quantities, in contrast to the host of all the atomic positions, generally change in a manner not reversible in time. This is well known from daily life, for the shape of a mug, as a collective quantity for all its constituents, changes dramatically when it breaks, without easily reattaining its original form. The temporal process of a jar breaking into pieces is clearly distinguishable from the pieces flying in miraculous formation, merging into an undamaged jar. One can easily distinguish a movie of the breaking from its temporal reversal, something impossible to do for the molecular motion.

The contradiction between the temporal reversibility of the motion of the microscopic constituents of a mug on one hand, and the macroscopic irreversibility of the whole mug's motion on the other, finds its explanation in the special form of the initial conditions required for a specific collective quantity such as the shape. Falling down toward the ground where it will break, the mug's constituents have highly coordinated initial positions and velocities. The broken mug, by contrast, is splintered into many single parts whose positions, orientations, and velocities are all very different from one another. Only with much difficulty could one set them in motion so as to recombine into a whole vessel in a temporal reversion of the breaking.

One major difference between an intact and a broken mug is the state of order. In an unbroken mug, all constituents as well as their velocities are, by their very construction, precisely arranged and related to one another. In a broken mug, the order is partially eliminated along with cohesion; splinters can lie arbitrarily in space. An intact mug is in a much more special state than a broken one, there being many more configurations to be recognized in a broken vessel than in an intact one. The physical measure of this kind of disorder is called *entropy*, defined by the number of microscopic states corresponding to the numberless atomic positions in a macroscopic state specified by a few collective quantities. Entropy is low for an intact vessel, but high for a broken one; and it normally grows larger much more easily than smaller. The jar was created in a state of low entropy, only to break at a later time, or at least become worn and inevitably reach a state of higher entropy.

But how can one manage to construct a jar of low entropy in the first place? The material must be formed in a special way, and stabilized in a process that lowers the entropy. This is possible only by using energy, gained for instance by burning fuel such as coal or oil. Burning itself is a process whereby entropy is increased, as an ordered solid or liquid substance whose constituents must to a certain degree move in a coordinated way is transformed into an unordered gas of higher entropy. The entropy of the fuel must be increased in order to fabricate a vessel of lower entropy. As one can show on general grounds, entropy production through burning always exceeds the loss of entropy from producing an ordered piece. Viewing all ingredients, entropy indeed cannot decrease.

In this way, an imbalance between past and future arises in our coarse perception, which disregards microscopic details. Our memory is based on what is perceived, and so it shows us a world that distinguishes the past from the future. And even if we could directly observe atomic motion, there would again be by far too much information to be recorded completely; it would have to be reduced to a few collective quantities in our memory. Thus the contrast between microscopic reversibility and collective irreversibility can be explained.

From the cosmological perspective, this leads to an oft-discussed problem: If entropy is ever-increasing, a long time ago it must have been much less than it is now. At one time, it should have had the minimally possible value, in a state of perfect order. How could such a state have emerged—one capable of starting the world but also dooming it never to be perfect?

We formulated this problem by applying our earthly experience to the whole universe. But in doing so, we overlooked a crucial ingredient: gravity, which is unavoidable in the cosmos. Gravity is able to increase order by virtue of its attractive nature, just one of its many special properties. If we view a uniformly distributed gas, such as matter as it was shortly after the big bang (as seen via the cosmic microwave background), it has decidedly low order and structure and thus relatively high entropy. Without gravity, that distribution is not going to change much; entropy does not change and remains at its high level.

But gravity inevitably acts and slowly leads to a concentration of overdense regions where chance (embodied by quantum fluctuations) has made the gas denser in the first place. Over time, ordered structures emerge in the universe and culminate in the formation of galaxies and stars. The entirety of all these structures is much more highly ordered than the initial state; by gravity's action, entropy was diminished. Every once in a while stars form, shining their light on a planet to continuously provide it with energy from the nuclear fusion taking place in their gravitationally compressed and heated-up interiors. Under suitable conditions, such as those on Earth, much more strongly ordered structures can arise, maintaining themselves as life-forms through consumption of the donated energy.

A better picture than ever-increasing entropy is indeed, as has been admitted by its initiator Ludwig Boltzmann, a nearly constant level, locally reduced by some physical processes or just by mere chance. In these oases of order, further processes play out to slowly increase entropy and bring it back to the initial level. And yet, imagining the universe from the viewpoint of an external, metaphysical observer positioned not in the universe itself but overseeing everything in it and its entire temporal extension, the present state of cosmological parameters indeed seems to be one of enormously high entropy. There are, after all, very many possibilities for what a universe of a certain size, expansion rate, and acceleration could look like. All details of the exact distribution of galaxies, stars, and planets, or even of the atoms in them, are irrelevant in this context, and thus the present universe more closely resembles a vessel broken into the finest pieces than an intact, precisely constructed one. The universe as we happen to find it appears as but one of many possibilities, which could all have the same form of cosmological collective quantities, such as the cosmic background radiation or the distribution of galaxies on large scales. Seen as a whole, the universe does seem to be in agreement with our experience of everyday systems, which has accustomed us to a continuous increase in entropy.

Cosmologists often see a problem here, since one assumes that entropy must have been ever smaller the earlier back in time we go. This becomes especially problematic if the universe did not start a finite time ago but existed at all times—possibly in a long succession of repeating cycles, sometimes expanding and sometimes collapsing, as suggested if quantum gravity removes the big bang singularity and as described in more detail in the next chapter. The farther back a theory lays the initial point of the universe, the smaller entropy must have been then in order to have been able to grow continuously to its current value. One must postulate a very special and highly ordered initial state, whence a deterministic picture leads precisely to the state of our universe at present (a process called *fine-tuning* in the language of cosmologists). By itself, a postulate of this form may be acceptable; but if a special initial state is introduced, a good theory should also provide a convincing explanation for its origin. Thus the issue has changed somewhat: How can one explain the present, apparently so special state of the universe without already assuming a very special initial state?

This viewpoint, with its problematic nature, is very influential, but is often only the result of faulty analogies. What we want to explain in cosmology is, for one thing, the picture given by the universe on large scales, and also the possibility of our closer neighborhood in the solar system allowing our existence as a life-form. With this, we essentially have the kinds of observations made now and for the foreseeable future. Seen cosmologically, the present universe is of interest only as a macroscopic state given independently of its details, and whose form follows quite well from current cosmological theories.

Cosmological observations, such as the distribution of background radiation or of galaxies, determine a few parameters as collective quantities not influenced by the precise positions of single stars. For our existence, properties of the solar system, for instance the distance of Earth from the sun, are tremendously important, for other values would make Earth too cold or too hot for life as we know it. But this is only a limited number of parameters, all playing the role of collective quantities. Compatible with this macroscopic state is a multitude of microscopic states — states that, except for our solar system, differ from the macroscopic state revealed by cosmological observations only in details below the galactic scale. In cosmology, such details are of no importance, except for properties of our solar system when they are essential for the existence, for instance, of life.

For a superobserver outside the universe, the current exact configuration of galaxies and stars down to our solar system, but also of other ones in the universe, would seem highly special. He would conclude that it could only have come from a very finely tuned initial state, no matter at what time it is posited. But we are no such observer, which cannot even exist in the physical world. We are observers who must draw our conclusions, as daring as they may be, from a position within the universe. We view not only the macroscopic state of the universe in cosmology and its collective quantities, but also a part of the microscopic state in daily life on Earth by our mere existence. A sizable part of our observational data from the cosmological standpoint is, even if it can be grasped by collective quantities, partially a characterization of the microscopic state.

Thus we are not in a situation in which we would be interested only in macroscopic quantities, a situation in which entropy is an important parameter. We sit as it were on a splinter of the broken mug, and for us this splinter is at least as important as the whole thing. There are innumerable configurations of splinters, but the chances are very small that any one of them will take the particular form of the one where we happen to find ourselves. As observers in the interior of the system as the whole universe represents it, we follow selected microscopic details and suspend problems as they would arise from an increase in entropy. We see our neighborhood very precisely, a standpoint counting as microscopic for cosmological states. We see a very special configuration, but one that is simply microscopic and random. The splinter where we live just happens to be as we find it; nothing more is to be explained.

THE ARROW OF TIME:
A WORLD IN SHAMBLES

Every moment just comes to imply those that follow.

—JEAN-PAUL SARTRE, *Nausea*

The existence of an arrow of time is often seen as one of the great mysteries in physics, especially cosmology. Sometimes entropy is identified as the reason for the directedness of time, as it would cause the total disorder in the universe to increase and thus distinguish the past from the future. But when one considers the partially microscopic character of our observations in the universe, entropy ceases to be a candidate for explaining time's arrow. Entropy always depends on the degree of precision used for observations and then allowed for a theoretical description. This cannot possibly explain the everyday phenomenon that we are continuously being pushed ahead in time without ever being able to step back or just have a little rest. Otherwise, atomic precision of observations combined with a photographic memory, making accessible time-reversible microscopic physics, would allow us to go back in time—which is hardly imaginable.

Before going into more details we should, to avoid misunderstandings, first clarify what the arrow of time is supposed to mean and, perhaps more important, what it is not. In particular, the question of why one cannot single-handedly reverse the arrow of time is independent of the enticing fantasy of time travel. During time travel in the usual sense, whose (im)possibility shall not concern us here, time does not run backward; it merely proceeds differently for some people. A person or a group of people is enclosed in a bounded region that, after a process rarely specified further, arrives in the past of its neighborhood. Neither for external bystanders nor for the time travelers does a step back in time occur; rather one region of space is merely separated from the outer world and subject to a modified process of time. While time travelers do venture into their neighborhood's past, their own time keeps running forward. After all, they retain their memories of all their past, a part of which is now in the future of their neighborhood, and they do not grow younger, or even die when a time before their birth date is reached. The question of why time, in contrast to space, is a strict one-way street is independent of the question of whether time travel is possible.

Why is there no going back in time? When all is said and done, this question is based solely on a misunderstanding. The primary experience is motion, and we move neither in space nor in time, but in space-time. It makes no sense to speak of motion in space independently of time, or of motion in time independently of space. What constitutes motion is only a change of position in space during a given interval of time. Even before we grapple with relativity, we can say that space and time are unbreakably linked with each other. Motion itself would be impossible if we had to decide the direction in time as well as in space. Elementary time abstracted from the fact of motion—independently of whether life invents such clever things as memory—is not a rigid parameter but merely a way of specifying change and motion.

As described at the beginning of this chapter, in physics one often uses the picture of a process happening in mathematical time; in fact, this is how calculations of theoretical physics are most often organized. But what one really measures are relationships of some observables with respect to others. Time itself is measured by an apparatus such as a clock; a measurement process thus provides the relation of, for instance, the position of an object to the hand of a clock. Introducing time is an additional step, defined by convention. Nor in cosmology does one measure the expansion of the universe as a function of time, even though this is the form that solutions to general relativity initially take. One measures the escape velocity of a star by the redshift in its emission spectrum as well as its distance from us by comparison with standard candles. Using general relativity, this relation can be translated into a time dependence of the expansion of the universe, subsequently to be compared with the mathematical solutions. Precisely this comparison has provided indications of the previously discussed existence of dark energy by the observation of many supernovae. As in this example, a relational description of motion, in which the ratio of some quantities to others is given and directly compared with observations, more realistically reflects what is actually being measured. This concept plays a role especially in quantum gravity. But it does not provide a solution for the directedness of time, because space is also relational and accessed only by distance measurements between different material objects. The relational behavior thus cannot explain the apparent difference between space and time concerning their directedness.

It may sound surprising, but the only alternative to directed time, which would make space and time play no different roles, is the impossibility of motion of any kind. Having time but being able to decide arbitrarily between stepping toward the future or the past is impossible; one would literally be paralyzed, unable to choose. For what would be the reference for a step back in time? What happened to me five minutes ago is the influence of the world (including my own body) on myself, when I was in a certain state characterizing my age five minutes ago. In different states, other acts occur, perceived by us as the flow of time since the mind is influenced by its own body to partially memorize past states. Time itself is an abstracted quantity: It is a parameter, in its special form chosen conventionally and used to determine motion; motion itself is fundamentally relational and describes a complex interplay between different things, not between a thing and time. In the same way, the psychological perception of time as the organizing principle of memory contents is ultimately grounded on an evolutionarily buttressed convention whose destruction would render impossible any recognition of motion. If a change of time's direction were to be allowed, this convention would be toppled. None of these relational states can be changed by conscious choice; but that would be required for a reversion of time subject to the will, truly constituting a return to past and potentially remembered states.

When attempting to explain the directedness of time, one is ulti-mately led to a problem much older than the relativistic theories of space and time: the problem of the existence of motion and change, a problem that was vexing philosophers as far back as Parmenides. He saw only one solution: a complete denial of motion, called by him pure illusion. Despite the obvious difficulty of reconciling this view with the most elementary observations, this hypothesis retained its influence through the centuries—starting with Parmenides' immedi-ate successor, Zeno, and the atomists Leucippus and Democritus, then via philosophers such as Schopenhauer to physicists like Schrödinger. The problem is still open from a philosophical view-point, but here we are already treading outside physics, whose task it is to *describe* change and motion, thus accepting them as already given.

A final chance to speculate about turning time's arrow around is sometimes seen in quantum cosmology. Again for mathematical rea-sons, it is often useful to take the volume of the universe as a parame-ter to describe temporal change relationally. This time surely progresses in an expanding universe, but what happens when the universe one day stops its expansion and then collapses? Its volume would start to decrease and develop backward—and with it perhaps time. This hypothesis was first postulated even before the advent of quantum cosmology, by Thomas Gold in 1958.

Such a behavior of the volume of the universe is easily possible in general relativity, even though current estimates of cosmological parameters, as they appear to be established for our universe, may not make it likely. Does time then turn around at this point of inver-sion, not by conscious decisions of observers in the universe, but by the evolution of the universe itself? The remote possibility of such a behavior may have to be granted, for the nature of time in quantum gravity is not yet finally clarified.

But it is much more likely that such an inversion of time's arrow would merely represent an illusion of the chosen mathematical description. As one wanders on Earth along a track across the North Pole, the latitude first motions upward. Once we pass the pole, the latitude decreases, but that does not mean that one would now travel back along the old trail. The longitude, after all, takes a different value on the other side of the pole. In the same way, the turning point of the volume of the universe, when used as a parameter for time during the collapse, would not mean that time then runs backward. Analogously to the longitude, one would just have to use other parameters to keep separate the universe before and after its maximal extension. Quantum cosmology thus does not offer clearcut reasons for the reversion of time either.

All scenarios of doom had, as always, turned out to be false prophecies. True, the universal economy had nearly collapsed, but this prediction, given all the panic-mongering, was self-fulfilling. Long ago, the background radiation of gravitational waves had been measured out so sensitively that the composition of the universe was known precisely enough to compute its future. After the acceleration caused by the transient dark energy had ceased, the expansion itself began to slow down and was now on the verge of stopping, setting the whole universe on a collapsing path. The exact time for the expansion's end was not known, but every estimate resulted in a value within several decades. At first it did not appear threatening, for it would take further trillions of years for the universe to become too dense and hot for survival. On the contrary, a return to a warmer universe appeared tremendously welcome.

And yet, voices that raised this turning point of expansion to the status of the end of the world grew louder. Despite all the successes of stoop theory, one question had remained unanswered: What exactly is time, and is its progress reversible or perhaps tied to the development of the universe? Old theories had related the direction of time to the expansion of the universe, even postulating a redirection of time's arrow in a collapsing universe. One would then remember the future and predict the past. This expectation was founded on some forms of quantum cosmology whose implications in such a big universe, as it was constituted then, were normally minute; all other progress notwithstanding, such theories thus remained poorly tested. Now people vaguely recalled the distant possibility of the reversion of the arrow of time—an idea that, most likely thanks to its esoteric character, spread fast.

Under those circumstances, it did indeed appear reasonable to squander all one's savings (at least within the limits allowed by the life processes, artificially slowed down for survival in a cold, empty universe). If the direction of time will soon revert, savings, interest, investments, and funds lose their meaning. Instead, one should spend all one's money and live life to the fullest, a luxury to be revisited after time's reversion. Many followed this principle in expectation of the collapse. Large consumer spending led to an economic boom never seen before. But when at last the expansion of the universe stopped and collapse ensued, no reversion of time's arrow occurred! The only cosmological consequence was that the universe now grew, slowly but unstoppably, warmer and ever denser—fuller: The stream of time was swelling . . .

8. COSMOGONY
MYTHS AND METHODOLOGY

It is not *how* things are in the world that is mystical, but *that* it exists.

—LUDWIG WITTGENSTEIN, *Tractatus logico-philosophicus*

The economy now was knocked to its knees; society was in disarray, for all resources lay in the hands of the few who had resisted the call to squandering. The civilizations wasted away, waiting for their demise in the heat of a cosmos becoming ever narrower. Ever so slowly the economy recovered, until after several billion years the upturn was finally guaranteed thanks to the surplus of energy in a universe once again warmed up. Technology and the economy, adapted to the old drought, nearly boiled over and produced unknown riches. Now, finally, the civilizations were able to use to the fullest the knowledge developed in the turbulent but long-ago expansion of the world's youth, then saved through the critical collapse. The few daily inconveniences were easily taken care of by highly sophisticated machines, and sufficient energy was readily available. A golden age commenced in which humankind was all but free of work, with culture and science blossoming to unforeseen extents.

All this happened in recognition of the finiteness and perishability of all life in the shrinking universe, as people decided to use their remaining resources as a cultural buoy, in the possibly pretentious opinion that their knowledge could be of importance to the universe—or something in it. Cultural achievements as well as molecular and anatomical details of the surviving civilizations were, following the example of the legendary capsule Kruskal, encoded in a strong *gravitational wave signal to save, if not life, at least the thoughts through the dense universe—as long as time itself existed . . .*

Once theory pushes forward to a possible understanding of the big bang and the remaining universe, the temptation to explain the emergence of the universe itself becomes overwhelming. Interpretations of theories and their mathematical solutions concerning entire worldviews indeed offer a high degree of fascination. But in too direct and supposedly generally valid an interpretation there lies, especially in this case, a great danger—not least because theories relevant for such questions will for all foreseeable time remain in their infancy. Physics is, after all, even if we disregard its big sister philosophy, not alone in this business. And yet a comparison of different worldviews offers a certain charm, and certainly some knowledge, too.

One should not underestimate myths and what they can teach us about ourselves and the progress we have made. Take the Summer Palace in modern Beijing (figure 30), a beautiful sprawling park built as the summer retreat of Empress Dowager Cixi. On a small island in a man-made lake, facing the Tower of Buddhist Incense and the Sea of Wisdom Temple on the slope of Longevity Hill which rises from the shore, stands the Hall of Embracing the Universe. It is a small, humble building in the style of its time, the fringes of its roof rising optimistically upward to aim at the sky. The Hall of Embracing the Universe tells us everything there is to know about humanity and the world: It was initially called the Hall of Watching the Moon Toad to honor its role in observing the moonrise; nowadays, the Hall of Embracing the Universe is a souvenir shop.

Surprisingly often, one can find parallels between ideas stemming from the most diverse traditions, an observation probably not hinting at an ember of truth but rather traceable back to the fact that the range of human imagination is, despite its excesses, actually quite small. Analogies listed in this chapter as examples are not at all intended to suggest strict relationships between the different approaches; for all their superficial similarities, these ideas differ in their details, let alone their messages.

Hall of Embracing the Universe (Hanxu Tang)

Originally a three-storey building named "Hall for Watching the Moon Toad" completed in Emperor Qianlong's reign (1736-1795), the hall was rebuilt with just one floor during the reign of Emperor Jiaqing. It became a wonderful place for emperors and empresses to enjoy the moonlight during summer and autumn evenings. From here Emperor Qianlong watched naval maneuvers of the Tough and Sharp Regiment of the Fragrant Hills. Burned down by the Anglo-French Allied Forces in 1860, the hall was rebuilt during Emperor Guangxu's reign (1875-1908). Empress Dowager Cixi also sat here to watch Naval Academy maneuvers.

民生人寿 敬赠
MINSHENG LIFE

30. The Hall of Embracing the Universe in Beijing's Summer Palace.

ANALOGIES:
DECEPTIVE SIMILARITIES

Analogies prove nothing, but they can make one feel more at home.

—SIGMUND FREUD, *Introduction to Psychoanalysis*

In several ways, scenarios for the emergence of the world have a role similar to that of the atomic picture of matter. As is well known, the name and the concept of atoms was introduced 2,500 years ago by the pre-Socratic philosophers Leucippus and Democritus, to be verified by physics only in the early twentieth century—not least by contributions from Einstein in the same year, 1905, when he published his theory of special relativity and the photon hypothesis. In both cases, the modern and ancient atomic pictures, bodies not further subdivisible, the atoms, are postulated as the elementary building blocks of matter. But details of and motivations for the pre-Socratic theory of atoms strongly differed from the physical concept founded on observations, even if one disregards the ultimate divisibility of physical atoms as demonstrated more recently. The atomists developed their concept as an answer to Parmenides' shocking assertion that all motion must be illusion. Parmenides founded this apparently absurd thesis on a few logical steps, based on the initial statement that nothingness does not exist. Then it follows that no motion can exist either, for that would require bodies to take positions where earlier the nonexistent nothingness would have been. This logical contradiction, maintained after Parmenides especially by Zeno, was countered by the atomists by simply and radically accepting empty space as a given, where atoms would move and form matter.

This philosophically impressive series of developments differs markedly from the emergence of the concept of atoms in modern physics. Nowadays, we are used to representations and direct images, for instance those of scanning tunneling microscopy, that, indistinctly yet unmistakably, show the construction of matter from single atoms. In physics, however, the concept of atoms was established much earlier, based not on direct but on indirect observations. Einstein, in one of his famous works of 1905, quantitatively attributed Brownian motion—a microscopic trembling of small particles such as pollen grains suspended in a liquid—to irregular bumping by molecules of the liquid. He not only raised this bold thesis, but was able to support it mathematically by providing an equation for the relationship of the rate of trembling to properties of the liquid particles. In the following years, precise and successful comparisons with observations quickly made the concept of atoms widely accepted,

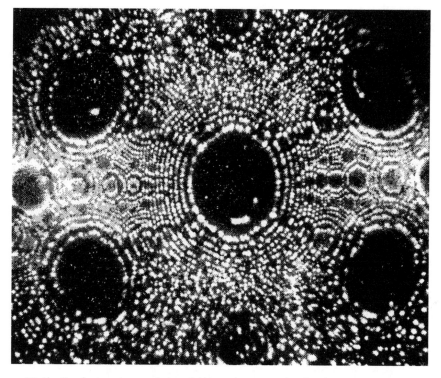

31. The tip of a tungsten needle in a field ion microscope, showing the atomic structure.

but it was another fifty years before Erwin Müller from Pennsylvania State University managed to produce the first direct images of atomic resolution using his own invention, field ion microscopy (figure 31).

With pictures of the emergence of the world, the situation is similar. Many proposals have been made throughout human cultural history, and they have often been garnished to the minutest detail. For some time, physics has been contributing to the resolution of such questions, and its methods have consistently been crowned by success: sensitive, though sometimes indirect, observations paired with a strong foundation of theory. Compared to the development of the concept of atoms, the phase we are presently in lacks a direct image; and it is questionable whether a direct image of the world's own emergence will ever be available. What comes closest to an image right now is the picture of the cosmic background radiation from times shortly after the big bang, but a direct image of the big bang itself, or even of the previous universe, remains a fantasy owing to the density as well as the quantum theoretical properties of this phase. Still, just as indirect observations of Brownian motion secured the physical concept of atoms, it may be possible to peruse sensitive measurements of the cosmic background of microwaves and other forms of radiation to gain insights into the ancient history of the universe. We have not come that far yet, but testability in principle allows us to use current theories for speculations.

MYTHS:
UNIVERSAL LIFE

People go there after they die!

—GIRL FROM JOSHIMATH

When classifying creation myths of the world, we can distinguish between primary and secondary creation; an illustration is given in figure 32. Primary creation provides a reason for the emergence of the world itself, secondary creation one for the world as we find it now. These may coincide, but they can also be separate.

Primary and secondary creation are identical for linear worldviews, which start with a moment of creation and progress continuously from then on, possibly to culminate in an apocalyptic endpoint. This scenario is well known from the Christian tradition, but it can also be found in the cultures of China.

More widespread is a separation of the concepts of primary and secondary creation, usually resulting in cyclic worldviews. This is realized especially in Hinduism: The world was primarily created by Brahma, an event followed by several cycles of destruction and re-creation. Destruction and creation are inseparably tied to each other, and combined in the gods' powers. An example is the explanation of why the god Ganesha has an elephant's head (figure 33).[1] Ganesha is

32. Primary and secondary creation: The act of primary creation is the artist's idea as well as its realization. Once primary creation is achieved, secondary creation takes place much more easily and almost automatically (here, by scraping the white wall above a judiciously placed black marble plate), which leads to the formation of structure as in the universe. (*Dispositivo per creare spazio* [*Device to create space*], 2007. Idea and realization: Gianni Caravaggio. Photograph: Paul Andriesse.)

33. Ganesha with one intact and one broken tusk.

the son of Shiva and Parvati, both gods, making for a rather unusual family life. One day, while Shiva is away for an extended period, Parvati delivers Ganesha, who quickly grows up into a strong young god. Returning home, Shiva sees this unknown youth, who has been entrusted by Parvati with safeguarding the house. Ganesha courageously blocks Shiva's way to his own house, only to be beheaded in the ensuing struggle. Having arrived on the scene too late, Parvati can do nothing but tell Shiva that he has just mutilated his own son and will not be allowed to enter the house before correcting the mistake. Shiva looks around and the first creature he sees is an elephant, whose head he puts on Ganesha's shoulders. (Even as a god, Shiva can apparently not travel back in time to avoid the fight; this may perhaps be taken as supporting evidence for our discussions in the context of time's arrow.)

In addition to such picturesque illustrations of godlike powers, one can find extremely detailed descriptions of the temporal progress of cycles as they occur between one secondary creation and the subsequent destruction. Even quantitative durations of cycles are given in years, whose basis is unfortunately unknown. In contrast to such ornamentations typical of the Indian culture, one symbol, the swastika, is distinguished by its simplicity and even abstractness. As the stylized form of a wheel, it symbolizes the cyclic progress of the world and also the arrow of time by the direction of the hooks.

The connection between a cyclic worldview and reincarnation or, more generally, an earthly existence after death is well-known. The following anecdote is illustrative: After a conference near the town of Roorkee, about a hundred kilometers north of Delhi, two Indian students and I had decided to use the remaining three days before our departure for a trip into the near Himalayas. After some inquiries among the local scientists familiar with the region, we started with a vague idea of a route along the Ganges. With the rudimentary public transportation, on a mountain road often only roughly built, the tour did not proceed very smoothly. But eventually we reached, after a day and a half, the hill station called Joshimath, not far from Gangotri, the glacier that sources the Ganges. At first, we took a ropeway up to where one has a view of the gigantic surrounding mountain ridges. But before the day ended, we also wanted to climb down the valley to the Ganges, which here—compared to its form beyond the Himalayas—is but a small though sometimes wild mountain creek.

We returned to the village and crossed it on a trail snaking down the slope. We could not see the Ganges from there, for it had, as it were, hidden itself behind the valley's horizon. Walking down the slope, cut through the millennia into a convexly curved V, wanderers cannot see the river, but only a stretch of the mountainside yet to be descended. The surroundings often remind one of the Alps, though soon enough the exotic impressions return. Higher up, we had walked through a forest of large trees densely covered by lichens, through deep snow softened by the March sun, white butterflies merrily flapping aloft. In the valley we strolled past lone cows ruminating by the trail and passed groups of children who just briefly interrupted their play to watch the (for them) exotic Westerner.

Among them we noticed a group of girls in colorful clothes. They were chatting beside the trail, on the roof of a building attached to the slope beneath. One of my acquaintances, Rakesh, turned toward the eldest of them to inquire about the route to the river—as usual, in Hindi, for English is not often spoken in this region—and so I was unable to follow the seemingly cheerful exchange. Afterward, Rakesh recounted the surprising conversation: The girl could not understand why we wanted to go down to the river, and upon further questioning, she explained: "People go there after they die!"

The meaning of this exclamation was obscure not only to me but

34. The Ganges Valley at Joshimath.

also to my two fellow travelers versed in Hinduism. Was it merely a prank played on strangers? Or a horror story told by parents to prevent their children from making a potentially hazardous descent to the river? But would this not even encourage the adventurous? Following the trail, we kept thinking about the girl's statement, so puzzling but uttered so sincerely. All the while, the river remained behind the slope horizon. Dusk came on, and the clouds began to cover the peaks for the night. Rain was forecast for later that evening and we were unsure whether the roads early the next morning would be negotiable by buses to bring us, after a weary eighteen hours, back on time to Delhi, the starting point of trains or flights to our dispersed homes. The darkness made us turn around having seen here in the valley nothing of that Indian Styx. Just as hidden remained for us the precise meaning of what the girl from Joshimath had communicated. But it symbolizes cyclic ideas of the world in the form of an existence on earth after death, closely interwoven with what is locally presented by nature.

Many other cultures also favor a cyclic worldview, not altogether surprising if one considers the primarily cyclic form of most phenomena, such as the course of the day, and their strong impact on nature and on cultures young and old. The changing day and the course of the sun played an especially significant role in ancient Egypt, where the sun god rode through the heavens as across a sea and every night had to tackle severe dangers that threatened the next day's rise. So it was not at all ruled out that daily cycles might one day cease as in a singularity. In Egyptian cosmology, too, there was held to be the "first time" of a primary creation, followed by infinitely many cycles. Moreover, Egyptian culture strongly relied on support from periodic floods of the Nile. Such vital events can easily find their way into an abstracted worldview. Examples also exist on the American continent, for instance among the Mayans and Aztecs, who arranged their calendars in disk shape, making the cyclic course obvious (figure 35).

In the early days of European culture, cyclic concepts played a crucial role, too—in spite of occasional appeals to Christian teachings, which preferred a linear picture of cosmogony. Western philosophy and science were initiated and subsequently given important momentum in pre-Socratic Greece. In Hesiod's tales, one finds the four ages of the world, an obviously cyclic worldview. In addition,

35. A Mayan calendar in disk shape.

one can see there traces of a discrete conception of time, especially in the representations of a sickle-wielding Kronos, whose every cut shortens the threads of life by a fixed amount, as in figure 36.

Such myths influenced pre-Socratic philosophy when it launched efforts to develop a logically founded worldview. We already saw the static picture of Parmenides, a view neither linear nor cyclic owing to the complete impossibility of any kind of change. Among the other pre-Socratic philosophers, however, one can find on offer almost every conceivable worldview. Since we are leaving the age of myths here and entering the dawn of the philosophical-scientific occident, we will cover pre-Socratic ideas in the next section.

Occasionally one comes across the comment that differences between primarily linear and cyclic worldviews correlate with the social structure of cultures. Cyclic worldviews are supposed to correspond with a tendency toward fatalism, since a cyclic progress of time would not reward special efforts. This eternal recurrence also worried Nietzsche—for other, gloomier reasons. Linear concepts, by contrast, are sometimes associated with the encouragement of innovation. This interpretation can be seen in news stories, for instance about the Indian software industry, which skillfully uses existing technologies but supposedly does not lend itself easily to new developments. I first heard this notion from a self-critical Indian colleague who, however, can himself serve as a counterexample to

36. Kronos (or Saturn), shown with a sickle symbolizing the discrete progress of time. In the background, the cross is part of the astronomical symbol of the planet Saturn, represented by a sickle.

this overly broad generalization, easily rebutted with examples suggesting a contrary conclusion. In Europe, for instance, the all-important beginnings of philosophy, which continue to influence intellectual discourse to the present day, were accompanied by primarily cyclic worldviews, while later on the Christian linear tradition was unable to counteract the paralysis of scientific progress in the Middle Ages, having in fact given rise to it in the first place.

THEORIES:
THE WORLD IN ONE'S HEAD

Awake! (not Greece—She *is* awake!)

—LORD BYRON

In ancient times, myths can be distinguished from philosophical or scientific theories only with difficulty. With Thales, for instance (about 580 years before the beginning of the Common Era), one finds the idea of water as the fundamental element, distinguishing it from the other three. The importance of this idea lies not so much in its own content but rather in having initiated a long chain of rapidly refined and mutually stimulating hypotheses among later philosophers. In retrospect, one can find surprisingly many ingredients of modern cosmological worldviews in this impressive list of ideas.

PRE-SOCRATIC PHILOSOPHY: THE BIRTH PAINS OF COSMOLOGY

As early as Anaximander, just a few years after Thales, one recognizes an immense step in abstraction: The origin of the world is founded not in a concrete element but in "the unlimited" (*apeiron*). Moreover, one can see here concepts of symmetry, for the unlimited as origin is, for Anaximander, also structureless and, by logical consequence, in equilibrium. Only in the course of world events do structures form, but in the end the world returns to its unlimited initial state: "The origin of all things is the unlimited. But wherefrom is their birth, thereto goes their death by necessity."[2] This formulation provides many rough elements of a cyclic worldview as in modern quantum cosmology.

The role of symmetry—realized here by the initial state being structureless—is important as the reason for strong simplifications in solutions of general behavior. Aristotle says about Anaximander's idea of the structureless: "For a thing sitting in the center (of the universe) with symmetric relations to the outer boundaries has no reason to move upward, downward or (maybe) sideways. And since it cannot move in opposite directions at the same time, it must remain at rest." This line of reasoning is strongly reminiscent of the symmetry arguments often used in theoretical physics. One could, for instance, solve a complicated differential equation with a homogeneous, structureless initial condition, and in the end find that the solution is motionless. However, it is much simpler to recognize the symmetry and, like Anaximander and Aristotle, to see no reason for motion from the outset. Such symmetry arguments, often in more complicated forms, play a large role in physics.

About seventy years after Thales, *becoming* is at the center of Heraclitus' philosophy: "I see nothing but becoming. Don't let yourselves be fooled! It is your short sight, not the essence of things, that makes you believe you see solid land in the seas of becoming and demise." Modern cosmology was led to envision the large-scale change of the universe only by direct observations of its expansion, as evidenced by the escape velocities of stars in distant galaxies. Even Einstein was initially misled by too short a focus and attempted to find static solutions of constant volume in his general relativity. He managed to do so by artificially introducing an extra term, the so-called *cosmological constant,* only to dispose of it soon after Edwin Hubble's discovery of the expansion of the universe. Only in recent times does the cosmological constant again play a large (though unpopular) role in the possible modeling of dark energy.

For Heraclitus, moreover, the worldview is cyclic in the strict sense—based on several cycles; and like Thales, he views one of the four ancient elements as the fundamental one. For him this is fire, which periodically destroys worlds and allows them to be resurrected in conflagration: "This world order, the same for all beings, has been created by no God and neither by Man, but it was always and is and will be forever lively fire, measuredly glowing and measuredly quenching." Heraclitus thus counts as a founder of *ekpyrosis,* later popular with the Stoics. Ekpyrosis has itself been resurrected in modern cosmology—by the same name—in a string-theory-motivated model by Justin Khoury, Burt Ovrut, Paul Steinhardt, and Neil Turok.

Parmenides, as we've seen before, is diametrically opposed to the insight of his contemporary, Heraclitus: All change is illusion. Nothing emerges or wanes, it simply is. The reason is of purely logical nature, for no being can come from nonexistence: "The decision about this relies in the following: it is or it is not! . . . How, then, could all that exists emerge in the future, how could it once have emerged? For if it did emerge it is not, and just the same if in the future it should one day emerge. Thus emergence is quenched and demise is lost." Emergence from nothing is impossible: "For unspeakable and unthinkable is how it could have been nonexisting. What duty should have driven it to sooner or later begin from nothing, and to grow? Thus it must either be at any rate, or not at all." In physics, in the same way, the beginning and end of everything exist at most in the form of singularities, outside any theory and logic.

Fifty years farther on, Empedocles follows in Parmenides' steps, but adds new elements. The relationship is clear, for instance in the following sentence: "For as it is impossible for something to emerge from the nowhere existing, so it is undoable and unheard of that the being could ever be eradicated." Building on this idea, a clearly cyclic run of the world ensues: "Now, insofar as in this way the one emerges from the many, and the many in turn sprout from the decay of the one, to this extent emergence happens, and life does not remain unchanged; but since their continuing change never ceases, they always remain untrembling [gods] during the cycle."

Crucially new with Empedocles is the insight that interplay of two counteracting forces is necessary for stable equilibrium: "For how [those two forces (conflict and harmony)] once were, so they will [ever] be, and I believe that never will the infinite eternity be deprived of the two of them." In his use of the terms "harmony" and "conflict" one clearly sees how Empedocles came to this insight. From the human domain, he quickly extrapolates to cosmology and grasps a crucial ingredient of modern physics.

Just before the time of Empedocles, forty years after Heraclitus and Parmenides, Anaxagoras had contributed a theory of the formation of structure. The initial state is homogeneous, as for Anaximander: "All things were together, infinite according to amount and smallness. For the smallness was simply infinite. And so long as all things were together, due to the smallness nothing could clearly be discerned therein." The world thus starts from a highly symmetrical and extremely simple initial state with little structure. As in modern cosmology, the crucial problem is how, from such a symmetrical state, structures such as galaxies and stars can arise; one must find a motivating reason for a complicated dynamic happening. For Anaxagoras, this reason is the spirit: "And over everything just having a soul, large as small, the spirit rules. Thus it has the power over the whole vortex motion so that it can push on this motion. And first this vortex began from a certain little point, but it extends further and will extend still more. And all that was mixing and separating and departing from each other was known to the spirit. And all ordered the spirit, how it should become in the future and how [what no longer exists once] was and how it is [right now]."

Except for the strong determinism, this picture closely resembles the inflationary formation of structure, assuming an initial structureless vacuum state whence structures emerge by quantum fluctuations. For the emergence of structure, quantum theory is crucial, replacing the "spirit" of Anaxagoras. It may be noteworthy in this context that some physicists, such as Eugene Wigner, have attributed an important role to consciousness in the measurement process of quantum mechanics, namely, in the collapse of the wave function. In this view, the wave function collapses when consciousness (of an experimenter or an observer) acts on the quantum system.

Nowadays, however, this opinion is shared only by a minority, and especially in cosmology it is dubious. In final consequence it would, after all, mean that the wave function of cosmic background radiation, which resulted from the initial vacuum, collapsed just a few decades ago when Arno Penzias and Robert Wilson measured them by a lucky break. (Unless, of course, there are other life-forms in the universe who scooped humankind in this discovery and thus caused the collapse even earlier.) But by preventing the gravitational clumping of smaller concentrations, cosmic background radiation plays an important role in the early universe, which was much denser than it is today. Without it, galaxies of much smaller scale than observed should exist. If the radiation at those times had still been in its original uncollapsed wave function, it would have interacted with matter differently, changing the process of structure formation. If consciousness is required for the collapse of the wave function, one can only draw the conclusion that in those early times a self-conscious matter form must have made observations—an assumption not at all justified to save Wigner's interpretation.

In addition to pictures of the cosmos, the consequence of indivisibility derived by Parmenides from the same logical principles as the illusion of motion is of interest: "Nor is it divisible because it is totally uniform. And nowhere is there a stronger being which could prevent its connectivity, nor a weaker one; it is rather entirely filled by the being. Thus it is fully connected; for one being closely borders to the other." As already seen, Zeno tried to support the hypotheses of Parmenides, but the contradiction with observations of obvious change was too dramatic.

This dilemma motivated the atomists, in particular Leucippus (a student of Zeno, who himself was a student of Parmenides) and Democritus (slightly younger than Leucippus), to question Parmenides' premises systematically. Motion obviously must be possible, and so the atomists eventually concluded that what exists is, after all, divisible. If that is the case, then being borders on nonbeing and motion becomes possible. In modern terms, this phenomenon is called the movement of atoms in empty space.

In this fashion the pre-Socratics anticipated many a question in cosmology that was made answerable only millennia later by modern physics, for instance by the work of Einstein on Brownian motion. But one step the pre-Socratic philosophers had not yet undertaken was to question the unlimited divisibility of space and time themselves. For this, the radical change of view engendered by general relativity was required, attributing a physical role to space-time.

Otherwise, one can find among the pre-Socratics most of the elements of modern cosmology. Only with quantum gravity did truly new elements enter the game. For instance, it poses the possibility of limits to the divisibility of space and time on a scientific basis, and from it derives cyclic worldviews by preventing the singularities of general relativity. The large-scale happening is determined by interplays between the classical attraction of gravity and quantum gravitational repulsion at small volume. As a qualitative feature this consequence, compared to pre-Socratic pictures, is not at all new. What is new here is the atomic structure of space and time as the physical cause; this was not discussed by the pre-Socratics. Above all, quantum gravity makes possible explicit, ever more detailed calculations; they not only further embellish the general picture and may one day make it empirically testable, but occasionally lead to the discovery of completely new phenomena. The most striking example of an essentially new ingredient is perhaps *cosmic forgetfulness,* leading to a mixture of linear and cyclic worldviews in which the end of every cycle is seen as something like a new beginning: a world foam-born in quantum fluctuations, a clean, fresh, virgin slate rather than the charred, barren, forlorn wasteland of Heraclitus' conflagration.

PHYSICAL COSMOLOGY: SOCIETAL CYCLES

Science entered into the business of cosmic worldviews relatively late, induced by nearly simultaneous progress in the observations of distant stars and galaxies as well as in the theoretical foundation presented by general relativity. Just after Einstein's equations were formulated, Einstein himself, followed by Willem de Sitter in 1917 and later Alexander Friedmann and Georges Lemaître, found simple solutions describing the temporal evolution of a universe that is isotropic on large scales. At that time, nothing was known of the expansion of the universe, and so scientists were primarily looking for solutions with a potential to correspond to a static collection of masses. But curiously enough, this was extremely difficult, and it became possible only with Einstein's improvisational introduction of an additional term in the equations whose size is determined by the cosmological constant.

At about the same time, astronomers such as Vesto Slipher started to notice that most of the stellar objects called *nebulae*—in contrast to stars, slightly diffuse—appeared to be moving away from us. One did not see the motion itself, but the nebulae's light systematically showed an increased redshift compared to that of stars. In relativity, this is traced back to an outward motion of the emitting object; here we have the analogue of the Doppler effect in acoustics, which makes the siren of a departing ambulance sound different from that of an approaching one or one at rest. Similarly, the frequency of a departing light source moves toward the red end of the spectrum. The reason for such a fleeing motion of nebulae was, however, unclear, for why should there be forces acting only on the nebulae but not on other stars? And why, one may innocently ask, should those forces move the nebulae away from us?

In 1929 the riddle was solved by Edwin Hubble, who was able to use precise astronomical measurements to determine the distances of nebulae from Earth. As in more recent observations of the acceler- ated expansion by means of supernovae, he perused clearly identi- fied standard candles whose real brightness—that which one would perceive near the star—can be derived from other properties accessi- ble from afar. Since the brightness as seen from Earth decreases with the distance to the object, the difference between real and observed brightness allows one to compute the distance. For Hubble, stan- dard candles were not supernovae but Cepheids: variable stars for which Henrietta Leavitt had noticed a close relationship between their brightness and pulsation rates. Hubble found numerous Cepheids in some of the known nebulae, and from their pulsation rate he could determine first the real brightness and then the distance of the host nebulae—with the exceedingly surprising finding that nebulae lie far outside the Milky Way. Thus, nebulae are diffuse sim- ply for the reason that they themselves consist of innumerable but very distant stars. Nebulae are nothing but galaxies of their own. By his observations, Hubble extended the understanding of the cosmos far beyond the borders of the Milky Way.

Moreover, the puzzle of escape velocities was now solved by the finding that more distant objects, independent of their kind, move away more rapidly than closer ones. Quantitatively, Hubble found a linear relationship between the escape velocity and the distance, a simple proportionality. The unflattering conclusion that we are so unpopular in the universe that not even galaxies can stand to be around us can easily be evaded by assuming a uniform expansion of the entire universe, of space itself. Galaxies do not move away from us or any other point; space between them and us expands. At a given moment in time, each piece of a line between Earth and a galaxy is increased by the same factor, making the change of the total length proportional to the distance. Moreover, observations found easy qualitative agreement with the cosmological solutions of general relativity—and that without assuming a cosmological constant. (Only very precise observations during recent years had revealed, as already mentioned, an acceleration of the expansion caused by dark energy. This again requires a special, déjà-vu contribution behaving in a way similar to that of a cosmological constant.) One could trust the theory, even when it was applied over such large cosmological distances. Here lies the birthplace of modern cosmology.

Alas, the solutions had a grave flaw: As one followed them backward in time, all of them gave infinitely high values for the matter density of the universe a finite time ago, a time when all of space had withered down to a single point. At that point, the infamous singularity, Einstein's equations lose all their sense; the point itself falls outside the physics described by the theory. And yet attempts have often been made to interpret the singularity as the starting point of the universe, as it were a scientific proof for a linear worldview. But based solely on the classical theory, such bold conclusions are not allowed; the singularity clearly shows that the theory cannot be applied at that point.

Only an extension of the theory can help here, rendering it still able to describe expanding solutions but without running into a singularity. Only such a theory can show whether the instant of the classical singularity can play the role of an initial point, or whether physical time extends beyond the singularity—before the big bang. Since its early days, quantum theory has been seen as a crucial component of such an extension of general relativity; but this enterprise could be tackled only very slowly because of substantial mathematical complications. Early versions of quantum cosmology gave pictures of how a nonsingular beginning could appear, but properties of the singularity as a temporal boundary of space-time were not questioned; physicists merely attempted to cover up, as it were, the singular behavior behind the uncertainty of quantum theory. Such ideas were developed in the 1980s, in particular by Jim Hartle and Stephen Hawking on one side and Alex Vilenkin on the other with different models, which will be described in more detail in the following chapter.

A series of partially cyclic views was proposed independently, with a certain behavior of matter or quantum theory postulated but not derived to be active during the big bang phase. In the 1930s, while the Great Depression reminded everyone that economic growth cannot go on forever, Richard Tolman was the first physicist to investigate the possibility of cyclic universes, with expansion followed by contraction and so on. To the present day, his studies, especially of the role of growing disorder and entropy in a cyclic universe, inspire ongoing research. After Tolman and the Great Depression, the ensuing decades—the 1940s through 1960s—were the heyday of Keynesianism, the theory that government intervention can spur economies to be ever expanding. During those linear years, the expanding big bang model was established in cosmology, with first results on cosmic background radiation published by Alpher and Herman and its eventual discovery by Penzias and Wilson. Hopes of an endlessly expanding economy were dashed by the oil crisis of the 1970s; but in 1979, physicists including Mario Novello and José Salím, and independently V. Melnikov and S. Orlov, proposed concrete mechanisms by which the universe could bounce back after a cosmic depression. Those ideas were used and extended in the work of Ruth Durrer and Joachim Laukenmann as well as Patrick Peter and Nelson Pinto-Neto in the 1990s and onward, providing entire cosmic scenarios in which structure can emerge without need for the negative pressure that must be present in an inflationary phase. Most of these investigations, however, were confined to the simplest isotropic universes and may, due to the required special properties, not prove valid generally. String theory has also found ways to construct cyclic pictures, as propagated by Gabriele Veneziano and more recently by Paul Steinhardt and Neil Turok, without, however, yet being fully supported. At the time of this writing, strong interest is again being shown in those fateful cyclic models—and again the world is trembling in financial crisis.

With the rise of loop quantum gravity, a new and different possibility arose to tackle this problem. Though it is still incomplete, we now had a quantum theory of gravity independent of the assumption of cosmic isotropy. Finding concrete solutions remained extremely complicated, but loop quantum cosmology soon made it possible, starting in 2000, to find the consequences of quantum theory for cosmological models. As already described, this indeed eliminates the big bang singularity by specific effects such as the discreteness of time as is typical for a combination of general relativity and quantum theory. To be sure, the desire to eliminate the singularity did not mark the beginning of loop quantum gravity, or of loop quantum cosmology either. It resulted from a long chain of mathematical constructions and calculations that made the nonsingular behavior visible only after several years of research. It is a consequence of the theory, seen independently of possible culturally influenced prejudices on the part of the participating researchers.[3] Here, as so often is the case, the unbribable nature of mathematics is of crucial importance.

More details and intuitive aspects of what replaces the classical singularity, as we attempt to bail out a crashing universe, can be found in the earlier chapter on loop quantum cosmology. Now, some of the possibilities of this mechanism will be introduced to show how it may, as its details are further investigated, lead to new worldviews extending the classical big bang model to a prehistory before the big bang. It is not yet clear which of the diverse phenomena will dominate the process, or in what ratios all the different quantum effects arise; some of them have been analyzed by Ghanashyam Date with Kinjal Banerjee and Golam Hossain, others with crucial contributions from Aureliano Skirzewski. No complete picture is yet available, but several scenarios have already been discussed.

A world model requires an understanding of the long-term behavior of the universe. It depends not only on quantum gravity in the big bang phase but also on parameters such as the matter content and the exact value of average spatial curvature. According to general relativity, these numbers determine whether an expanding universe can reach a point of maximal extension and then contract, or whether it must continue to expand for all eternity. Cosmologists tend to favor the first case—on the one hand, to avoid the dreadful idea of a universe ever more diluting, cooling, and thus increasingly dreary, but also because a cosmos initially contracting from infinite size, then rebounding once in a strong quantum phase, and finally forever expanding may be more difficult to explain. (But see the pre-Socratics for possible philosophical advantages of this *apeiron* picture.)

In a universe that is expanding, recollapsing, and rebounding time and again one has, with finitely long cycles in the infinity of time, more wiggle room to make preceding cycles responsible for the existence of the special conditions we find now. In the multitude of all cycles, very different and life-threatening conditions could often exist. But it suffices if in the infinite set of all cycles, every so often a part of a universe resembling ours can come about. As long as this is possible, with infinitely many attempts it is bound to happen; we would get rid of all need to explain why the universe is as we see it. However, Zeno's ghost reappears once more, as infinity is abused to make the existence of a world like ours plausible. As an explanation for the world, this is hardly satisfactory.

For an example of a model crucially relying on the passage of many cycles, we can look at the so-called *emergent universe*, proposed by George Ellis and Roy Maartens in 2004,[4] initially on a purely classical basis. According to the emergent model, the universe starts in a state resembling Einstein's originally constructed static world model, but it differs slightly in some properties so that it is not entirely independent of time. In the emergent model, starting near the static universe, there is a series of cycles, all very short-lived. The model contains matter in such a form that some of its properties change slightly during each cycle, eventually assuming the negative pressure required for an inflationary phase. With inflation, the ensuing cycle can expand much more than its predecessors and thus in principle resemble the part of the universe visible to us now. And for some specific properties, certain constraints arise: The picture is in principle testable by comparison with observations.

There are a few problems with the classical model. For one, Ellis and Maartens must somehow avoid the singularities of general relativity, possible only by specialized constructions. Then, the static universe is unstable, making a start of the universe near this state very unlikely. It is like throwing a ball onto a corrugated surface such as an egg carton and asking where one would most likely find the ball. It would certainly not be on the elevated parts, which constitute unstable positions, but rather in the valleys. Even near a peak one would not be likely to find the ball, for it would soon roll down from there. The static Einstein universe, as a consequence of its instability, corresponds to a hill, and thus the original emergent model has severe difficulties in reliably explaining the precise initial condition it requires.

LOOP QUANTUM COSMOLOGY: COMING FULL CIRCLE

Interestingly, both problems are elegantly solved by a combination with loop quantum cosmology, as George Ellis has shown with David Mulryne, Jim Lidsey, and Reza Tavakol.[5] Singularities are avoided in any case by quantum repulsive forces, easily providing a cyclic model. But independently—and more surprisingly, if one already knows the singularity avoidance—a new static universe, of much smaller size than Einstein's, arises; and this one is stable. It is an ideal starting point for a universe that initially, an infinite amount of time ago, spends its time in a very simple form only to go out and develop ever more in innumerable cycles. In this way, it gives rise to more pronounced complexity and eventually finds conditions suitable for inflation, kick-starting our part of the universe; mathematical solutions are illustrated in figure 37. As the third contribution from loop quantum cosmology, counterforces to attractive gravity, which drive the expansion, make it easier to manage inflation.

There are still more scenarios in loop quantum cosmology, but they do not crucially differ from what is sketched here. Common to all of them is the cyclic nature of the universe; linear forms such as

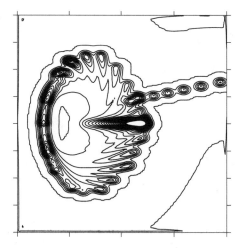

37. Illustration of a universe starting with continuously growing cycles and eventually exploding in an inflationary phase. The vertical position corresponds to the expansion rate (upper half) or contraction rate (lower half), the horizontal position to the size of the universe. Contour lines encode the probability of finding a state of a given change rate and spatial size in the history of the universe. The starting point lies within the enclosed oval region, whence the cycles describe a spiral growing outward. The final inflationary phase is realized by the ribbon protruding to the right.

appear in older versions of quantum cosmology do not come into play. Loop quantum cosmology always provides time before the big bang, but not a starting point of the universe a finite time ago.[6] It seems clearly partial to the latter view in the question of whether physical principles favor a linear or a cyclic worldview. But it also brings in an entirely new feature, something to be taken into account for further detailed formulations of worldviews: *cosmic forgetfulness*.

For many quantities, such as the total size of the universe or its expansion and contraction rate, the universe follows a deterministic process, a unique history free of singularities. But if we ask whether, in principle, all properties of the preceding cycle can be determined from observations, we will be disappointed: There are properties, such as the size of quantum fluctuations, that cannot practically be reconstructed from observations conducted in the next cycle. Herein lies the new element that surprisingly arises in loop quantum cosmology. Thus the picture is not purely cyclic, but contains a linear component; for the nonreconstructible properties acquire, as it were, a fresh start after every bang, even if other properties can traverse it in full remembrance. Mathematically derived details of quantum cosmology provide new principles, in this case a mixture of cyclic and linear worldviews, that in spite of all fantasy have not surfaced before. The further buildup of these pictures, especially with regard to possible observations, is currently in a phase of rapid progress.

She now called herself Quman[7] — with no small touch of unaccustomed irony, but maybe also in a whim of spontaneous gratitude. She rooted her new name in history, relying on deciphered messages from the human age of her predecessor universe. But a name was actually not needed, for she is a collective quantum form of life, based on sensitive superpositions in the wave function of the universe.

With growing density in a contracting universe, quantum mechanical properties gained more and more importance, and relieved the streaklike Quman from her former isolation in the all too classic phase. Now she blossomed, in the spring of a new universe:

Once more unfolding, fully correlated: world-comprising
And strongly heated, feasting, to excited high states rising;
In quantum acts of measurement herself entangled feeling,
And so with wildest quantum jumps into the future heeling.

Her existence is threatened only by decoherence, continuously gnawing away at her entanglement. But as a contemplative quantum form of life, she can decrease its implications by, so to speak, propagating by self-observation, thus producing copies of herself that follow her in time but, because of quantum uncertainty, are never completely identical.

Freed from individuality of any kind and thus from territorial thinking, she passes a peaceful existence; her survival in decoherence relies indeed on her collectivity. And yet decoherence had almost doomed this age-old culture in the preceding universe. In an unsettling episode, the world had swung itself up to an unusual size, assuming very classical and strongly diluted features in which entanglement of the wave function could persist only in very small regions. At that time, the Quman was finally unable to withstand decoherence even though quantum processes, and the simultaneous processing, as in a quantum computer, was programmed in her very existence.

In the end, she saw only one way to save herself: projecting her entanglement, at least partially, into what an intermediate life-form had once grandiosely called its "consciousness." But even here she seemed sentenced to perpetual slumber in a large universe apparently doomed to expand ever more, becoming ever more classical. Large quantum fluctuations no longer occurred that at other times had reliably caused a recollapse, a reversal of the expansion to contraction. By lucky (or conscious?) choice she had projected herself into an extremely aggressive life-form; in a suicidal move, these unwitting creatures would finally cause a recollapse. With their last breaths, in a whim of greatness, they had even encoded their cultural achievements, and unbeknownst to them, with these a part of the collective consciousness, in classical and quantum fluctuations of the cycle's latest stages.

As always, reconstruction against cosmic forgetfulness after the big bang was difficult. But Quman had success thanks to an extremely quick translation into her own entanglement just before the universe could swing back up to larger volume, erasing the last traces of old quantum fluctuations. Still, a large part of the dramatic previous phase was irrecoverably forgotten, and so Quman led a new beginning in cosmic history. Even though she did not remember it, this must indeed have happened many times before, whenever a cycle had become too classical and she had to yield to decoherence. But instinctively she knew and accepted it, undeterred in expectation of her eternal recurrence . . .

9. ONE WORLD
IDEAL SCIENCE

This world, now, is arranged precisely as it must be
to be just barely able to last: but would it be only a
bit worse, it could no longer persist.

—ARTHUR SCHOPENHAUER, *The World as Will and Representation*

Uniqueness is the strongest form of explanation. If a solution
is unique, it can be only so, not otherwise; it is clearly recognized as a consequence of the underlying theory and its
principles. In cosmology, uniqueness plays a special role, for we can,
after all, observe just one single universe: the one in which we happen to be situated. This can be called an observation (though an
extremely elementary one), to be explained by an encompassing
physical theory. Uniqueness here is not just an ideal or a mathematical exercise; it is a necessity for the scientific identity of cosmologists.

Examples of statements of uniqueness are many. A well-known
version in philosophy is Gottfried Leibniz's characterization of the
world as the best of all possible ones. This is an explanation of the
world based on uniqueness, if one assumes (with Leibniz) that
the goodness of worlds possesses a unique maximum. Schopenhauer
reversed the arguments, holding that our world should be the worst
of all possible ones, and he even believed he had a proof: namely, that
if it were even just slightly worse, it would, and humankind along
with it, have perished long ago. As one can see, Schopenhauer had
presaged the problems of the twentieth and twenty-first centuries.

INITIAL VALUE PROBLEMS:
A MOMENT OF TRUTH

Uniqueness, alas, never comes without cost. As in the two philo-sophical examples, assumptions are always necessary; they may appear more or less natural, but they must nonetheless be posed. In physics, they arise in two very different forms: assumptions needed to build up a theory, and assumptions by which solutions are selected within a given theory. The former is much more difficult to analyze and will be the topic of the next chapter. Selecting solutions, by contrast, is a standardized operation, tightly connected with the form of most mathematical equations as they are used in formulating physical laws. As repeatedly mentioned, these are dif-ferential, and sometimes difference, equations, giving the change of a quantity when one moves in space or time (or in another, abstract parameter).

But for a unique solution it is not sufficient to know only its rate of change; one also needs a firm and fixed starting point whence the changes start. If we had a good model for the development of stock prices, telling us by a mathematical equation how fast the rates go up or down at any given time, we would have to put in the starting val-ues at one time. Similarly, in physics, such a start could be an initial condition (one would say that the studied quantity begins with a certain value at a chosen moment in time) or a boundary condition (one would say that the quantity takes given values at the boundary of a spatial region). Since relativistic physics knows space-time only as a single object, not space and time in separation, one can combine these two kinds of conditions into a single one. (The structure of dif-ferential equations suggests a distinction also in the relativistic case. But distinguishing space and time does not play a role in this chapter, in contrast to our earlier discussion of time's arrow.)

Initial or boundary conditions can be seen as the theoretical equivalent of an experimenter's decisions in constructing and performing an experiment. The theory itself, by contrast, should, at least approximately, correspond to the behavior of Nature under the natural laws imposed on her. An experiment is always a special situation in nature specified by the experimental setup (for instance, a pendulum) and the initial configuration (the position in which one lets the pendulum swing freely). Such a selection is theoretically described by boundary and initial conditions. In other words, the theory used is selected by choosing a specific natural phenomenon, and the conditions for its solutions stem from the specific realization of the phenomenon within its general possibilities.

In cosmology, the first point—the general phenomenon of the universe—ultimately leads to quantum gravity as the theory for its description. But then there is only one specific realization of the phenomenon: our universe. A selection of initial or boundary conditions should not at all be necessary; instead one should, by cold logic and in consistent keeping with the general circumstances, expect a unique solution of the theory without fixing any further conditions whatsoever. The problem now is that in quantum gravity, as in other areas of physics, we are dealing with differential or difference equations—again, equations for change, starting from a fixed initial point to be chosen in addition. These equations find all these multifaceted applications in physics because they represent such a powerful method that hardly any other mathematical construction can eclipse them. They also determine what happens in cosmology, as has been impressively confirmed by observations, but as usual they require the specification of a starting point for solutions of the universe in addition to the equations themselves.

Will the aim of explaining the uniqueness of our universe remain a fantasy? The statement of our universe's uniqueness may fall dangerously close to a tautology, but the desire to derive it by physical methods is surely reasonable. We observe the world in its details and can never see the universe as a whole, but a large part of physics consists of extrapolations, sometimes exceedingly daring ones. Why should we not try to abstract our small-scale knowledge of the cosmos to obtain a theory extrapolated to the whole universe, possibly proving the uniqueness of the world? This undertaking represents a colossal test not only of quantum gravity but of the identity of physics. The little that is known about such questions will be introduced in the course of this chapter.

THE WAVE FUNCTION OF THE UNIVERSE:
GRASPING EVERYTHING

At first glance, quantum cosmology may look like a harmless application of quantum mechanical rules to the whole universe—especially if the cosmological principle of homogeneity is appealed to, making one consider only the total volume but no finer details. Quantum mechanics was originally developed for atomic physics, but it soon turned out to be a much wider framework that could encompass all physical processes. Not only are emission spectra of atoms explained by quantum mechanics, but also the rich and various phenomena of solids, for instance, electric conductivity in metals. Even astrophysical objects, most importantly white dwarfs and neutron stars, can be understood only by means of quantum physics. From here, applying quantum theory to the whole universe is not too long a step away.

On closer view, however, this ultimate step does have a special flavor. Here, at the outermost frontier of scientific knowledge, the remarkable but rather strange properties of the quantum mechanical wave function become a severe problem. After all, the wave function is not directly observable but instead, as it were, keeps a record of all information possibly gained from measurements on the system it describes—with all the limitations arising from uncertainty relations. It is an administrator with built-in ignorance. In the interpretation accepted in the course of time, the wave function describes accessible properties of the quantum system as it was prepared by an experimenter, who now begins to undertake measurements from outside the system.

Such a viewpoint can be applied without any problems even to "large" systems such as a piece of metal or a whole neutron star. A metal physicist does not stand in the metal, and an astronomer is far away from a neutron star. A cosmologist, however, cannot be separated from the object studied: the universe. In contrast to all other examples of quantum physics, quantum cosmology must always grapple with an observer who is necessarily a part of the investigated system. It is impossible to split quantum cosmology into a wave function on the one hand and an observer measuring its properties on the other. There is only a wave function of the universe, supposed to describe everything including us as observers (or theorists).

In full generality, all this information is certainly impossible to compute, and one must rely on extreme simplifications and approximations to make any progress at all with such questions. But in all these models, one can never avoid eventually dealing with the peculiarity of an observer within the system—an all-encompassing wave function that, in a sense, measures itself. One can occasionally find a possible escape route in the suggestion to assume a superordinate observer who would powerfully watch the whole universe and its wave function, all at once and from outside. One could then use the usual rules of quantum theory at least without mathematical problems; but even so one does not fundamentally avoid the question of what we—as observers in the system, not as external observers—can measure about the wave function of the universe.

While of general nature, these questions play a role especially for the problem of uniqueness and quantum gravity. For the desire for a unique solution starts from the fact that we as observers in the universe can see just one universe. Were we to pose this question for a superordinate observer, we would have no indication whatsoever how many universes could be seen. If an imaginary superordinate observer can see our universe as a whole, why not others, too, or even all possible ones? Although such observers are sometimes assumed in physical investigations—mostly as a last (cheap) refuge from daunting conceptual difficulties—they have no relevance for physical questions. Physically, only what we can perceive ourselves, in our own universe, can play a role. In quantum cosmology, there is no avoiding this issue.

ONE UNIVERSE—NONE UNIVERSE: THE BABY IN THE BATHTUB

How to interpret the wave function of the universe is already one question pertinent to quantum cosmology, but no other quantum system. Perhaps, then, it is possible to tackle the wave function's uniqueness in a special way. Even quantum cosmology relies on the same type of mathematical equations as the rest of physics, with initial values at a fixed time to be selected for a unique solution. Whether the time selected is indeed the absolute beginning of the world, or merely the beginning of some phase under theoretical consideration, a decision is, apparently, always to be made.

The first proposal to deal with this problem in quantum cosmology was put forward by its founder, Bryce DeWitt, as early as 1967. Aiming to connect the question of uniqueness with the singularity problem, he postulated that the wave function of the universe should be zero for a space of vanishing volume. According to general relativity, the spatial volume is zero just at the singularity, and so DeWitt's condition corresponds to an interpretation of the singularity as a beginning where one would, as elsewhere in physics, fix initial values.

But as a condition at the singularity, this procedure would be more powerful than using ordinary initial conditions: DeWitt here attempts to play the big problems of quantum cosmology—the classical singularity and uniqueness—against each other. At first, his condition in a sense eliminates the singularity in quantum cosmology. If the wave function vanishes there, the universe will, according to the interpretation of the wave function, never assume the state of the singularity. This would be a consequence of the condition chosen for the wave function, rather than a physical phenomenon such as a repulsive force in loop quantum cosmology. Still, if successfully implemented, it would have far-reaching importance. Further, independent implications, testable at least in principle, would result from the specific form of the wave function as the solution of a differential equation with DeWitt's initial condition.

Uniqueness of the wave function would be just such a consequence, important enough to promote the singularity avoidance achieved by DeWitt's condition to a wide and elegantly formulated principle. But there is a catch when different cosmological equations are considered. Indeed, DeWitt's condition often implies a unique wave function, naturally able to describe the oneness of the universe. In most cases, however—especially more realistic ones of less symmetry than in exactly isotropic models—this unique wave function vanishes not just at the singularity but everywhere. Here, the baby is thrown out with the bathwater: Such a universe would avoid not only the singularity, but any geometrical state—it would not exist at all. The wave function, after all, gives the likelihood of measurement results—here, of the size of the universe; if the wave function is completely zero, there is no possibility of measurements, and thus no universe either.

This disastrous contradiction with most elementary observations quickly led to the downfall of DeWitt's condition. Later, in 1991, Heinz-Dieter Conradi and Dieter Zeh made an attempt to prevent the failure of the condition based on postulated changes in the equations of the universe at small volume, modifications expected anyway from a general quantum theory of gravity in the form of quantum corrections. But without progress in the general development of quantum gravity, the situation turned out to be too complex to achieve the correct form by anything more than guessing. In the framework of loop quantum gravity, now available, we will shortly come back to this issue.

PHYSICAL PICTURES: NOTHINGNESS WITHOUT BORDERS

Two alternative approaches gained more popularity than DeWitt's condition: the tunneling condition of Alex Vilenkin and the no-boundary condition of Jim Hartle and Stephen Hawking. Both conditions, as different as they are and as heatedly as they have been debated, continue to put the beginning of the universe at the place of the classical singularity. In those cases, however, the wave function there differs from zero to avoid DeWitt's fatal problem of an entirely vanishing wave function.

Vilenkin's tunneling condition relies on another effect of quantum mechanics, again a consequence of properties of the wave function. A wave function can often penetrate barriers with its tails, even if those would be too high for a corresponding classical particle. Many a physical phenomenon is based on this possibility, such as some kinds of radioactive decay and several technological developments including new transistors in microelectronics and the scanning tunneling microscope. Boldly stretching the analogy, Vilenkin proposed in 1983 that the universe itself might have emerged by such a tunneling process. Our universe would be the tail of a pioneering wave function that had once penetrated the barrier of the big bang and its singularity. But from where did the universe tunnel, and from where came the bulk of the wave function, whose tail our universe is supposed to be, before the tunneling process? Vilenkin's answer, obvious only at first sight: From nothing.

One should perhaps, in general and especially in this case, not read too much into the notions of physical theories and grasp them as merely what they are: names that paraphrase a mathematical fact by analogies. Nothing cannot be thought; or, formulated more strongly, "Who thinks of nothing does not think at all":

SOCRATES: He then who sees some one thing, sees something which is?

THEAETETUS: Clearly.

SOCRATES: And he who hears anything, hears some one thing, and hears that which is?

THEAETETUS: Yes.

SOCRATES: And he who touches anything, touches something which is one and therefore is?

THEAETETUS: That again is true.

SOCRATES: And does not he who thinks, think some one thing?

THEAETETUS: Certainly.

SOCRATES: And does not he who thinks some one thing, think something which is?

THEAETETUS: I agree.

SOCRATES: Then he who thinks of that which is not, thinks of nothing?

THEAETETUS: Clearly.

SOCRATES: And he who thinks of nothing, does not think at all?

THEAETETUS: Obviously.

SOCRATES: Then no one can think that which is not, either as a self-existent substance or as a predicate of something else?

—**PLATO**, *Theaetetus*

One can hardly attribute physical meaning to tunneling from nothing in a literal sense. Regardless, Vilenkin's postulate does have sense with regard to the wave function of the universe, endowed by the tunneling condition with certain initial values at vanishing volume. There, the wave function is no longer zero, and it is not doomed to imply a wave function that vanishes everywhere. Instead, it prescribes the rate of change of the wave function at this place, a sufficient specification to determine it completely. One can try to draw at least rough predictions for the further course of an expanding universe at larger volume, and compare them with observations.

The no-boundary condition of Hartle and Hawking from the year 1984 sets up an initial condition in a similar way, motivated by a physical picture. Here, the picture is that of a universe closed and rounded off in the past, where classically the singularity would lie; consequently there is no boundary to the past—hence the condition's name. Also here, one can imagine that the classical singularity is replaced by this rounded-off space-time; the proposal thus combines the singularity problem with the question of uniqueness. It may at first appear problematic that general relativity does not allow any space-time closed off at its bottom in the past, like some kind of chalice. This is the crucial point where Hartle and Hawking postulate new equations deviating from general relativity. Since deviations occur only for a small universe near the singularity, they might well incorporate implications of quantum gravity. Like all initial conditions mentioned so far, however, this one does not arise as a derivation from an encompassing theory, which indeed was unavailable in those years, but rather relies on generalizations of quantum mechanics to cosmology.

DYNAMIC INITIAL CONDITIONS: STATELY LAWS

MANTO: I stand still; around me circles time.

—GOETHE, *Faust*

While a strict derivation of quantum cosmology from a complete theory of quantum gravity is still lacking, there are, by now, numerous indications of important phenomena. In a concrete way, loop quantum gravity provides equations extending those of general relativity. Though not yet fully formulated, they have been tested theoretically in a variety of ways. In this process, different corrections to the classical equations have crystallized and can be considered characteristic of such theories of quantum gravity. We have already seen implications for the elimination of singularities at the big bang as well as in black holes; now we shall turn to the question of uniqueness.

Equations of loop quantum gravity are not yet available in full generality; its effects can rather be seen in different models. An analysis tells us that indeed conditions on the wave function of the universe automatically arise. In their details, they are closest to a mixture of DeWitt's condition and the no-boundary condition of Hartle and Hawking. The problems of DeWitt's original condition are avoided since the equations deviate from those used by him owing to quantum corrections, including the important repulsive forces at small sizes of the universe. Loop quantum gravity thus offers a possibility to walk the path proposed by Conradi and Zeh in a systematic manner. The similarity to the no-boundary condition then shows up in the further course of the wave function farther away from the singularity, where its height increases in both cases. By describing the wave function as the tail of a tunneled universe, the tunneling condition, by contrast, would lead to a wave function of a different shape, its size decreasing at larger distances from the wave function's bulk situated in "nothing." Older proposals can thus clearly be distinguished from one another and made concrete.

In contrast to the old conditions, there are two crucial differences in loop quantum cosmology. First, the big bang singularity does not occur as a beginning or a boundary of the universe, but is merely a transitory phase—one of extremely high density and violence, but transient nonetheless. And yet this classically singular phase has implications for the wave function and is responsible for the constraints it is subject to. Second, these constraints are no longer imposed as physically motivated conditions; rather, they follow from the quantized Einstein equations themselves. Even though the exact form and strength of the conditions—whether there is always a unique wave function or possibly a constrained but larger class—is not yet clarified, an outlook to something very new in physics is nonetheless offered: a system whose theoretical description is not split into dynamic laws of nature and initial conditions in the hands of a physicist, but where initial conditions follow, as it were, dynamically, as a consequence of the laws. That alone would be the exemplar of a cosmology, the realization of the dream of an ultimate theory explaining not only the temporal course of the universe but also the fact that there is only a single universe.

Here, on the most elementary level of the wave function, we are led back once again to the singularity problem. In spite of the existence of repulsive forces to prevent the collapse of the universe into a singularity, a regular behavior of the universe's wave function is not yet guaranteed. Repulsive forces build up barriers that are too high for the universe to surpass in its collapse; instead it is, once it has collapsed, pushed back at a certain minimal extension to reexpand. This is the situation shown to us by effective forces of loop quantum cosmology. But a quantum mechanical wave function is rarely impressed by a barrier, through which it can simply tunnel. We have so far disregarded this problem because the main part of the wave function would obey the backward thrust at the barrier. If, however, some part can proceed to the singularity, it may yet bring down the whole theory. The crucial question for a final solution to the singularity problem in quantum gravity is whether the wave function can push forward all the way to the singularity. And if so, does it mean the end of quantum gravity as a complete description of the universe?

At this juncture the final fate of quantum gravity is to be decided: Will it constitute a complete and consistent theory extending general relativity; or will it remain singular, itself to be extended in some as yet unknown way?

Loop quantum cosmology here proceeds as follows, as I observed in 2001: First, it does allow the existence of a state with vanishing volume, where the classical theory would assume its singularity. It approaches the problem in an unbiased yet daring way; it does not, as attempted by DeWitt without success, pose the absence of a singular state as a condition from the outset. Then it lets the dynamic equations, used to describe how the wave function of the universe evolves, decide for themselves which role this state shall play: the apocalypse of a singularity, or merely a harmless transition point.

In simple systems containing the usual symmetry assumptions of cosmology—the cosmological principle—the mathematical equations lend themselves to relatively easy analysis. These are not differential but difference equations, as on page 118, realizing a time that is discrete rather than continuous. While a differential equation at each place provides the direction to be followed by a solution curve, as in figure 3 or figure 4, a difference equation determines steps by which a solution has to change in a fixed time interval. The astounding result: The solutions of these equations in loop quantum cosmology remain entirely uninfluenced by the value of the wave function at the singular state. Unflinchingly, the wave function of the universe wends its way before and behind the big bang, without even taking notice of the potential singularity. In particular, the temporal evolution does not break down. It leaves the singularity isolated, standing still, separated from the world's course of the wave function before and after it—or around it.

Decoupling the singularity from the evolution of the universe has a further consequence. The dynamics of loop quantum cosmology is provided by a collection of mathematical equations, one for each change of state when transiting from one discrete time step to the next. If one of the states—the singularity—is decoupled, one equation too many remains, unused for the evolution. It is to be solved nonetheless, and gives exactly the condition desired for a unique wave function.

In this way, initial conditions are dynamically imposed in loop quantum cosmology: They are not independent of dynamic equations—natural laws—but can be derived from them. Here we have the ultimate triumph of discrete time, for the decoupling of the singularity and the related constraints on the wave function would not occur if continuous time, as in old quantum cosmology, would allow one to reach arbitrarily closely to the singularity. Firmly tied to uniqueness is the final prevention of the singularity: Even if the wave function can tunnel through the mighty barrier of repulsive forces, this subversion does not cause the breakdown of the theory.

ONE WORLD?:
THE LAST IDEAL

What is the status of the wave function's uniqueness in more complicated situations that do not exactly exhibit the cosmological symmetries? A completely general rule for the uniqueness of the wave function of the universe, together with a powerful mathematical method to compute it, would surely have numerous applications by virtue of the predictions it would make. In jest, the physicist Murray Gell-Mann (who won the 1969 Nobel Prize in Physics for his contributions to the quark model of particle physics) posed it thus in a question to Jim Hartle: "If you know the wave function of the universe, why aren't you rich?" Hartle's sober reply: "One can certainly derive the course of stocks from the wave function of the universe. The result is that it will rise with 50 percent likelihood, and fall with 50 percent likelihood." This exchange illustrates the special behavior of the quantum theoretical wave function, which, although it characterizes a system completely, does not allow strictly deterministic predictions even if it is known fully.

Theoretical physics is a long way from infiltrating the stock markets, as seductive as such an enlargement of research funds would be. This is a consequence not least of the second part of our condition for an explicit applicability, namely, the strong control over the mathematical evaluation of a possibly unique wave function. Even if one could prove the uniqueness mathematically, a concrete computation would be too complex to provide predictions of the slightest use in everyday life. Large-scale predictions of cosmological interest would, however, be conceivable, and thus the uniqueness question is of importance. Intellectually, this outlook represents an opportunity just as enticing for an understanding of the universe. But even if we are concerned only with the clarification of uniqueness rather than its greedy exploitation, many questions remain open once we leave the range of the simplest models.

To describe the real world, many extensions of the currently understood models must be undertaken. Those models are *isotropic*—they look the same in all spatial directions—in contrast to the real world. Isotropy, realized around each point in space, comes combined with homogeneity. Such a universe looks the same at each place, clearly different from the real world. In loop quantum cosmology, one can formulate models including anisotropy and inhomogeneity. One then has to deal with a huge number of equations in which not only are changes of the size of the universe described in successive time steps, but also simultaneous spatial changes. When completely formulated even in compact notation, such equations occupy the space of several pages, and computers of currently available technologies would fail at a numerical solution.

In such situations, one must rely on abstract investigation independent of explicit solutions, something not unusual in theoretical physics and mathematics. But the question of uniqueness of solutions for such systems of equations remains complicated and is, unfortunately, far from being clarified. Certainly the decoupling of singular states still implies constraints on the wave function; but it is not guaranteed that this suffices for uniqueness, or perhaps that it may not be, as with DeWitt's initial attempt, too strong a restriction for the wave function's own good.

Of at least some help is the supporting fact that the decoupling of singular states, as well as the number of constraints on the wave function, is independent of the precise form of matter in the universe. These are, instead, pure effects of space-time geometry. A uniqueness analysis would not be impeded by open questions as to the exact material content of the universe, such as dark energy or the form of matter at the high energies of the big bang. To be sure, the exact form of the wave function does certainly depend on matter, causing the evolution of the quantum universe just as it does for classical space-time. But whether or not one has a unique solution is unaffected.

An interesting indication of the soundness of dynamic initial conditions exists in the form of a statement of consistency between cosmology and black holes. Black holes, after all, host a singularity, too, and they are described in quantum gravity by a wave function. As in cosmology, singular states decouple from the rest in the temporal evolution and so imply constraints. These conditions can be evaluated in the simplest model of a black hole, surrounded by no matter and not rotating—a black hole as it is described in general relativity by Schwarzschild's solution. The exterior of such a black hole—outside the horizon—is completely static: No temporal change whatsoever occurs. There is, after all, no matter to fall into the black hole, and the black hole itself, owing to the absence of rotation, sits still in space-time: As in Aristotle's description of Anaximander's structureless universe, quoted in the context of pre-Socratic philosophy, rotational symmetry deprives the black hole of any incentive to move or change; the surrounding space-time is thus left unbothered by temporal change.

In this case, the equations of loop quantum gravity cannot yet be formulated and analyzed in exterior space, but they can be inside the horizon, as Abhay Ashtekar and I endeavored to do in 2003. This is the basis of results for the form of black holes in a quantum theory of gravity. Interior equations, as well as the Penrose diagram in figure 29 based on them, show that the classical singularity is penetrated as in cosmology and that, according to current evidence, the interior reconnects with the exterior behind the singularity. Also as in quantum cosmology, constraints on the wave function of the black hole arise from the decoupling of singular states. The compatibility of those conditions in the interior with the static behavior outside is an important test of consistency for the form of black holes in quantum gravity.

As pointed out first by Daniel Cartin and Gaurav Khanna in 2006, those conditions merely imply that the wave function behind the singularity, in the classically invisible part, assumes the exact mirror image of the wave function before the singularity—in the classically visible part. As unexciting as this may seem, it shows the consistency of all evidence currently available for the behavior of black holes: We know that the singularity is penetrated by the wave function in the interior, and we know that the exterior of this kind of black hole is static. Indications make it plausible that the interior must be connected with the exterior before as well as after the singularity; there is, in other words, no splitting off into a daughter universe. Now, if the static exterior is subject to no temporal change, the interior can be connected with it only if it behaves behind the singularity just as it does before it (even though it is not static itself). Exactly this is required by the dynamic initial conditions; and all secure insights, as well as indications still to be buttressed, are tied to each other consistently.

If the black hole is not situated in empty space, the exterior space-time is far more dynamic and complicated, but also more interesting. A precise understanding of the connection between interior and exterior regions by means of the wave function and its constraints would be crucial for predictions of what happens after Hawking evaporation of a black hole, and what astrophysical consequences this might have. Studies of the wave function in cosmology as well as black hole physics thus promise deep insights for our understanding of the universe, even if a possibly unique wave function cannot be exploited financially.

In the human age—in the universe cycle before she began to call herself Quman—the life-forms, which in their consciousness sheltered Quman in the all too classical phase, once settled on the planet Earth. Here there was a parasite, the lancet liver fluke, that counted ants as well as cows among its hosts. To transit from an ant into a cow, the parasite wandered to the ant brain, influencing it in such a way that this host bit down tightly on the tip of a blade of grass and remained there, thus ensuring the transition into a grazing cow as the next host.

Quman had found refuge in the human brain, where she withstood the classical drought. Tied up as she was, she could only wait. But like the little lancet liver fluke, she patiently worked toward the transition into a new host—a new cycle of the universe in full quantum freshness. Divided among countless individuals, progress was often frustrating; but gradually it took place. As the parasite with the ants, she induced her host to engage in apparently senseless behavior. She made him so greedy that he destroyed his own habitat against all reason—all this with the aim of driving him into outer space. When Earth was dying, Man had just managed to jump off to other planets and their moons. Instead of learning from old mistakes—something he never did—Man once again bit down tightly on the outermost tip of a blade of grass: Again he destroyed his now extremely meager biotope and at last had to settle other star systems, then the whole galaxy. Here, Man, lacking other energy sources, learned the cultivation and manipulation of black holes. Thus he came to influence the universe itself and push it, as the last blade of grass, to its collapse and the renewed quantum bath for a resurrected Quman.

Was all this done with foresight, with Man as plaything of a more intelligent life-form? Quman does not follow any plan, any more than did the lancet liver fluke. She settles in the universe and accepts its every turn and swing. Worries she never has, for well she knows that all happens as is writ—in the wave function of the universe.

10. THEORY OF EVERYTHING?
PHYSICS AND HUBRIS

For if he once, by chance, uttered the most perfect thought, he would not know so himself. For only delusion is given to all.

—XENOPHANES OF KOLOPHON, Fragment

In the preceding chapter our topic was the status of unique solutions of a given theory and its laws. A whole other question is that of the uniqueness of the laws themselves. Since a unique theory would have to describe all that is observable in the world, it is often termed the "theory of everything."

PRINCIPLES:
SCIENCE FOUNDATION

Strictly speaking, there is no "presuppositionless" knowledge; the thought of such a thing is unthinkable, paralogical: A philosophy, a "faith" always has to be there first, for knowledge to win from it a direction, a meaning, a limit, a method, a *right* to exist.

—FRIEDRICH NIETZSCHE, *On the Genealogy of Morality*

The question of the uniqueness of a theory is of a very different quality from that of the uniqueness of solutions within a theory. For a set of equations, it may be difficult to decide about the uniqueness of its solutions, but the decision is based on a clear mathematical procedure. Physically, a unique solution, if it exists, has irrevocable meaning by a comparison of the properties it implies with observations. Whether a mathematically unique solution is also physically relevant can thus be tested in principle. But how does one define the uniqueness of a theory, and how would one perform physical tests to confirm the relevance of its uniqueness? That such questions have now come to be asked in physics is another sign of its immense progress, realized in large part thanks to quantum theories of gravity, in particular string theory.

In all aspects of uniqueness of a whole theory there is always a degree of arbitrariness. Constructions of theories most often start from physically motivated and general principles that one would like to see incorporated. General relativity, for instance, makes use of principles realized so successfully in special relativity, and tries to extend them to the gravitational force. General relativity is not the only possibility for accomplishing this; thus the theory is not unique in this sense. But among all comparable theories, it is the most successful in its agreement with observations, the most elegant in comparison to theories of similar success; it is therefore concretely elected by experiments, not by a purely mathematical uniqueness proof. Still, the use of the term "elegance"—a high degree of mathematical economy, for instance, in the length of resulting equations— already indicates that willfulness may enter the decision.

When desired principles become strong and precise enough, a mathematical treatment of the uniqueness question can come into reach. How, then, does one arrive at those principles, strong enough to put fantasy in a straitjacket? Since theoretical physics, like all science, grows historically, principles initially arise from analogies or generalizations of other known and already tested theories. Most principles are not of intuitive form that would evoke evident consent, having been strongly abstracted in historical developments. Abstraction surely simplifies a transfer to theories yet to be built, but it does not change the fact of the long period of growth through contributions from individual researchers.

Hence one can first infer the dependence of principles, as well as theories constructed upon them, on the existence of traditional streams in physics research. At its frontiers, after all, physics is never secure in the sense that all researchers involved would agree with all developments at all times. The emphasis on some principles as opposed to others is dependent on researchers' preferences. Willfulness can find here an especially easy entrance, particularly since concrete motivations arise not just purely intellectually, but all too often, unfortunately, from research-political considerations or even personal differences and vanity.[1] Moreover, inertia adds to this tendency: Once a decision on certain principles has been made, it is sometimes difficult to distance oneself from them even if their inadequacy has been recognized. Also, research life has accelerated; a reorientation would cost too much time and make one fall behind hopelessly in the competition for publications, research funds, and jobs. In the end, old principles are often defended for their own sake, and actual science is lost sight of.

These considerations mean that the uniqueness of theories is always conditional, for the underlying principles must first of all be accepted. Despite the claimed uniquenesses of theories, very different candidates with the same aim but different principles can easily coexist. Quantum gravity, with string theory, loop quantum gravity, and several other alternatives, is one example. While their shared ambition is clear, their foundations are not. Such theories are growing toward the end of their viability: They are plants whose seeds have fallen among the rubble, tasked to grow up toward the light. Especially when there is no glimmer of observation to show the way, this can be a long, roundabout way through a vast dark maze. Wander there too long, and the plant will become etiolated and may not survive to reach the light.

In order to bring within reach the question of a theory's uniqueness, a decisive mathematification is initially a strength. It allows a very precise formulation and a clear decision about uniqueness. To gain knowledge of nature, however, this process poses a disadvantage, for observations enter only indirectly via principles employed by the theory. Once a foundational decision about principles has been made to construct a theory and its mathematical apparatus, nature is no longer consulted in the uniqueness analysis. Even if a mathematically unique theory were to arise in this way, what would this mean for Nature ignored in the process? As quoted in the introduction, it is not up to physicists to impose laws on Nature, as mathematically elegant as they may seem.

Once mathematics is put to the task, allowed reformulations are always equivalences. The result does not mean less and it does not mean more than the assumptions, even though they always look surprisingly different. To exaggerate, all mathematical theorems are trivialities. The result is already contained in the assumptions, if often very veiled. In the unveiling, the high art of mathematicians shows itself, along with the importance of mathematical results in numerous applications. But by itself mathematics is no good as a model of nature. For this, observations remain necessary to crystallize a theory, be it a mathematically unique consequence of certain principles or not. No mathematics of a "theory of everything," however sophisticated, can replace this.

ONE THEORY, ONE SOLUTION?
BE CAREFUL WHAT YOU WISH FOR

Thou'rt like the spirit, thou dost comprehend.

—GOETHE, *Faust*

The ideal would be a unique theory with a unique solution. If only one could compute the solution, at least approximately, one would command complete control over all natural phenomena. Mathematically, something of this form would be quite conceivable—under the aforementioned qualification that assumptions are always required for the uniqueness of a theory. From a physical perspective, however, there is a different situation. As astonishing as it may sound, the possibility of a unique physical theory and one with a unique solution are not compatible with each other!

Suppose we had a theory with a unique solution. Given sufficient control over the form of its solution, we can directly test the theory by comparison of its unique predictions with observations; the theory, together with its solution, is scientifically testable. The number of observations, now, is always finite even though the sheer amount of data in modern physics has grown very large. Mathematically, by contrast, we can compute arbitrarily many properties from a solution—without any upper limit. At any given time, one can always think of new observational tests of the theory that have not yet been performed, tests for which the accuracy of available measurement technologies may not yet suffice. One can never completely verify the theory, but at most prove it false should any test not be passed[2]— a well-known fact in science that now, in the context of the uniqueness of theories or solutions, acquires new meaning.

Since we can never completely test a unique theory, we always have the option of slightly changing it, for instance by choosing different parameter values or delicately switching its underlying principles. If this is done with sufficient care, the uniqueness of its solution can be maintained, and one can remain in agreement with all experimental tests performed. With further physical progress, of course, an observation no longer compatible with every solution of all formerly possible theories would be made at some time in the future; some solutions will be ruled out. And if every one of the considered theories has a unique solution, whole theories would be falsified when their solutions failed. The wiggle room for differences in successful theories decreases as science goes on, but will never be restricted to just one possible theory. If one has a theory with a mathematically unique solution, the theory, understood in the physical sense, cannot be unique.

From the logical inversion of this statement it follows that a unique theory cannot have a unique solution, though we certainly have to keep in mind the fine differences of mathematical and physical uniqueness. Interestingly, recent developments seem to confirm this inference partially and in an astounding manner for a theory— string theory—for which mathematical uniqueness has been claimed. In the search for those solutions of this theory that potentially describe at least the simplest properties of experimentally known elementary particles, a whole landscape of slightly different solutions has opened up—a field of solutions of unimaginable size, whose number would dwarf that of all the protons in the universe. And every one of the solutions could be compatible with all experiments done so far. (Although this number is enormous, and damning to the uniqueness of the theory's predictions, just being able to estimate the number of solutions is an impressive feat.)

One is reminded of a precedent from philosophy, with morals taking the role of quantum gravity. Over the centuries, many philosophers had attempted to construct a unique theory of morals founded on clear principles. Best known among the examples is perhaps Kant's categorical imperative as a general principle whence individual rules of conduct should be derivable. Here we have the same problem as in string theory: The noble principles of theory allow innumerable down-to-earth solutions without being of real help in selecting a manageable set. Nietzsche recognized this most clearly:

> ... a vast new panorama opens up for him, a possibility makes him giddy, mistrust, suspicion and fear of every kind spring up, belief in morality, every kind of morality, wavers—finally, a new demand becomes articulate. So let us give voice to this *new demand*: we need a *critique* of moral values, *the value of these values should itself, for once, be examined* . . .[3]

The procedure in such a situation is, however, different for physicists than it was for moral theorists. While many theoretical physicists still keep up their hopes that further developments of the theory will once result in additional constraints on the solutions, strongly reducing the size of the solution space, others turn necessity into a virtue by simply declaring the investigation of this enormous set to be a new discipline. Physical predictions can no longer be obtained from a concrete solution? Let us then appeal to probability arguments. If we can only find sufficiently many solutions with a certain property, let us suggest this property also for our universe! Implicitly, we make an additional assumption: that our universe is typical among all those mathematically possible. This is again a physically untestable proposition, for we have but one universe, and what should we compare its type with? Unfortunately, many of the probability arguments, with their giant total number of solutions, come dangerously close to an act of Zenoic desperation:

Instead of dealing, as usual, with the concrete instances of the universe given and accessible to us, a great number of possible worlds is introduced. Problems of our world, solely in its theoretical description to be sure, disappear in a shoreless sea of utopias, never to be solved.

11. THE LIMITS OF SCIENCE AND THE NOBILITY OF NATURE

This noble metaphysical delusion is added to science as its instinct and time and again leads it to its limits, where it must turn into art: the actual aim of this mechanism.

—FRIEDRICH NIETZSCHE, *The Birth of Tragedy*

The idea (or fantasy?) of a theory of everything is ancient and powerful, and so it is not surprising that it often formed the background of secret societies such as the Pythagoreans or the Rosicrucians. It has entered science only recently, even as it may again take the appearance of a secret society due to the impenetrability of the subject. Sometimes, central statements of theories take the form of a dense web of interlocking (but unproven) conjectures, difficult to see through even for insiders. Especially in such cases, there is a considerable danger that the whole edifice will collapse like a house of cards, should someone only make a serious attempt to check it.

According to Pythagorean thought, unbroken integer numbers were deemed to take the role of elementary quantities. This surely is no theory of everything, but a guiding principle dominating further considerations. Should this hypothesis be correct, everything in the world must be expressible by integer numbers or their ratios (fractions). Building on this, the Pythagoreans developed impressive mathematical results, even though the precise attribution of individual bits (such as the "Pythagorean theorem") is historically uncertain. In spite of everything, the theory of the Pythagoreans had a fundamental flaw: Not everything can be expressed as a ratio of integers. For instance, the diagonal of a square with a side length of one meter has a length in meters represented by the square root of two: not a ratio of integers but an irrational number. Even in Pythagorean times this flaw was recognized—a shock from which that school could not recover despite apparent attempts to cover up the unfavorable result.

Hypotheses and unproven conjectures play a large role in science because they can, if sufficiently solid, stimulate further investigations. Even if they eventually turn out to be false, they crucially contribute to the progress of knowledge. It is, for instance, a truly masterly deed, one not to be underestimated, to prove the irrationality of the square root of two—or even just to question its representability as a fraction. The very emphasis placed on fixed principles often spurs critical scientists to disprove them. This is not at all meant to be distrustful (at least in most cases), but comes from the desire to secure the principles to the greatest extent possible. Therefore, sufficient openness must exist in diverse schools of thought, such as those investigating the theories of quantum gravity, so that their adherents listen to outsiders and give them access. The Pythagoreans did not pass this test of scientific scholarship and may instead have tried in vain to preserve their status by means of coverups. A revision of some principles in response to the new insights about the square root of two might have diminished their authority somewhat, but it would have allowed the continued existence of the school based on the valid parts of its theories.

During the same pre-Socratic times, an entirely different tradition ruled in cosmology and philosophy: one of open contests among small schools such as those of the philosophers mentioned in the chapter on cosmogony. Here the dominant viewpoint was that of critical rationalism: Theories were developed rationally but examined critically. Through these contests, a long line of highly innovative insights and cosmological worldviews was generated. In the quality of other results as well, such as the astronomical ones of Parmenides, these philosophers were not at all second to the Pythagoreans.

Such different traits can be found also in modern research, depending on the personalities of the researchers involved.[1] The all-too-human component of science, all objectivity notwithstanding, is not to be underestimated. Our basic perceptions and their process-ing mechanisms initially emerged through evolutionary adaptation to conditions on Earth; now they are being strained in an effort to understand the whole universe, on both small and large scales. The human mind has led to unexpected successes, in which mathematics as ordering power has played a decisive role. But can we be sure that this is not a wrong path, or that essential ingredients of the world are not being overlooked by our repurposed senses?

Science can never rule out the possibility of taking a wrong path, but it can reduce its likelihood. The approved means is the versatility of critical rationalism as it was lived by the pre-Socratics: It is dam-aging if the majority of scientists in one area, such as quantum grav-ity, work uncritically on the same methods. Diverse approaches must not only be admitted but supported actively, especially in cases in which no observations can yet indicate the correctness of the approaches taken. Too strong a formation of scientific subgroups— secret or open—is certainly a disadvantage, since individual research is too easily suppressed. At the end of science, or so one hopes, stands Truth. For this a strong guarantee exists thanks to observation and its ultimate objective confirmation by Nature, for Nature can-not be bribed (even though bribery attempts can be made). But only that truth can win that has been given a chance to compete.

Unfortunately, a strong focus on a single or few directions occurs all too easily in the current research situation: Once having arrived at an influential position, often determined by chance or fashion, a research direction can easily strengthen itself by attracting funds and influencing new hiring. Frontier areas of research are tenuously secured and are correspondingly occupied sparsely. Even slight fluc-tuations in the balance of forces between different approaches can have much further-reaching implications in these fledgling areas than in larger, more established scientific fields, where they can eas-ily be compensated for. From a research-political perspective, it is in fact in the interest of an upstart discipline to suppress the direct com-petition of alternative approaches. Too strong a drive to exclusion can be counteracted only by independent oversight committees, which serve to rule out intellectual nepotism.

Despite the almost intoxicating progress of science, one must always keep in mind its limitations, which become clear especially at its frontiers. We have seen examples: the wide landscape of solutions in string theory, sold as unique; or the possibility of cosmic forget-fulness in loop quantum cosmology, in spite of the theory's strength in explaining events before the big bang. Such limitations can possibly be surpassed with improved methods, but unless and until this has been accomplished, they have to be taken into account.

An undeserved privilege of humankind is to have reached such a pronounced understanding of nature at all. Limitations in no way diminish this achievement, but rather confirm what Immanuel Kant had already identified as the nobility of nature, starting with human nature, characterized as follows in his *Critique of Practical Reason:*

> This respect-inspiring idea of personality which sets before our eyes the sublimity of our nature (in its higher aspect), while at the same time it shows us the want of accord of our conduct with it and thereby strikes down self-conceit, is even natural to the commonest reason and easily observed.

By the end of his *Critique,* he is drawn to apply his insight to the whole of nature:

> The former view of a countless multitude of worlds annihilates as it were my importance as an animal creature, which after it has been for a short time provided with vital power, one knows not how, must again give back the matter of which it was formed to the planet it inhabits (a mere speck in the universe). . . . But though admiration and respect may excite to inquiry, they cannot supply the want of it. What, then, is to be done in order to enter on this in a useful manner and one adapted to the loftiness of the subject? Examples may serve in this as a warning and also for imitation . . . so as to prevent on the one hand the errors of a still crude untrained judgment, and on the other hand (what is far more necessary) the extravagances of genius, by which, as by the adepts of the philosopher's stone, without any methodical study or knowledge of nature, visionary treasures are promised and the true are thrown away.

Kant even attempts to found hereon the possibility of moral law in all regards. For cosmology, it is sufficient to correctly value the scope of scientific knowledge about the whole world and to appreciate it—in as broad a frame as warranted but no broader. Success often blinds, nowadays as before the advent of general relativity and quantum mechanics:

> Hubris today characterizes our whole attitude toward nature, our rape of nature with the help of machines and the completely unscrupulous inventiveness of technicians and engineers; hubris characterizes our attitude to God, or rather to some alleged spider of purpose and ethics which is lurking behind the great spider's web of causality.[2]

Recognizing the nobility of nature also demands a sense of the incompleteness of our description of it. How is this realized in quantum gravity? String theory impresses with its magic and its possibly unique mathematical formulation, but like the Mephistophelian sulfurous smell there remains a bitter aftertaste of a wide, vast, and disorderly landscape of solutions. And the dynamics of loop quantum gravity is so complicated that it probably repelled many a researcher at first sight, as the Earthly Spirit did Faust. Perhaps we glimpse the true nature of the world in one of these forms, but we do not know and may never find out.

So often in this context has an end of physics been proclaimed—in the positive sense of putatively complete knowledge, and in the negative one of incurable estrangement from reality. Throughout, one should bear in mind how far physics has come from its beginnings to its current state. New insights will be added, mistaken ones expunged. This process—not a complete dominance over nature, even if it may be understood "only" in the form of perpetually valid natural laws and not direct influence—is the aim of science. Here one acknowledges the incompleteness of our description, for that is unavoidable. Incompleteness of understanding in no way diminishes the sense of wonder about nature, which probably has moved most scientists to choose their profession and which motivates nonscientists to occupy themselves with scientific results.

In quantum gravity, impressive possibilities of new knowledge can be expected for the foreseeable future. Theoretical progress continues unceasingly, and some surprising gains have resulted from the research of recent years, bringing hope for a fundamental mathematical analysis that will reliably reveal new phenomena. Cosmology will be made ever more precise by a series of satellite and other observations arising from the continuous development of new technologies. Ever more precisely will we be able to look back to ever earlier times, and eventually we should be able to make physical measurements of quantum gravity effects.

How far back in time can we be guided in this way? Will we ever, with a precision that meets scientific standards, see the shape of the universe before the big bang? The answer to such questions remains open. We have a multitude of indications and mathematical models for what might have happened. A diverse set of results within quantum gravity has revealed different phenomena important for revealing what happened at the big bang. But for a reliable extrapolation, parameters would be required with a precision far out of the reach of current measurement accuracies. This does not, however, mean that it is impossible to answer questions about the complete prehistory of the universe. Cosmology as well as theoretical investigations are currently moving forward and will result in unforeseen insights. Among them might well be experimentally confirmed knowledge of the universe before the big bang.

NOTES

INTRODUCTION

1. Rudolf Carnap, *An Introduction to the Philosophy of Science* (New York: Dover Publications, 1995), 207.

1. GRAVITATION

1. Other striking examples are provided by the main theories covered in this book: general relativity and quantum mechanics. These theories started perhaps more controversially than Newton's, for they came at a much later stage and were tasked with explaining more subtle phenomena. In these as in other cases, it is almost shocking to realize that purposeful mathematical manipulations—pure thought—can so closely correspond with nature. So often, observations have been preempted by bold predictions, for instance of elementary particles envisioned before they were seen, following purely mathematical symmetries. Then again, theoretical visions so often turned out to be incorrect; but by being right much more than once, theoretical physics has abundantly shown that it did not just have a lucky strike. Its attempted predictions are indispensable in guiding observations and leading physics to new frontiers. Being granted confirmation of one's own ideas, even if they are rather minor, is every theorist's dream, though only few can have success. It is as if one thinks one is playing a game of chess, only to find oneself leading an actual battle.

2. The transformability of space and time does have consequences for daily life, for without it the amount of natural radioactivity on Earth's surface

would be much less: muons, elementary particles produced in the upper atmosphere by cosmic rays bombarding Earth, would otherwise decay high up in the atmosphere without being able to contribute to cosmic radiation on the ground.

3. The origin especially of the highly energetic contribution is not completely understood, but it appears to come from active galactic nuclei outside the Milky Way.

4. Such arguments may also be a telltale sign of the hidden agenda to conveniently avoid learning complicated theories, which only rarely find their way into standard university courses.

5. For this discovery, Hulse and Taylor were awarded the Nobel Prize in Physics in 1993.

6. "Heads" here literally refers to the body parts. If the minds of people in Europe and America also sometimes seem to point in different directions, that observation has an unrelated origin.

7. Accordingly, clocks slow down when brought to lower altitudes. This holds true only as long as the clock remains outside Earth's main material body, or within the light atmosphere. If a clock is dropped deep down into a borehole within Earth, the clock rate's slowdown is halted as a result of the distribution of matter now surrounding the clock from all sides.

8. Initially, the GPS system was developed for military purposes, and it is still overseen by the U.S. Department of Defense. Satellite signals available for civilian applications were, for some time, intentionally disturbed so as to allow position measurements only to within about one hundred meters. Over time, more and more, and very lucrative, applications were invented, and the disturbance mechanism was switched off in May 2007 following a directive from the Clinton administration.

9. Friedrich Nietzsche, *On the Genealogy of Morality.*

10. Undoubtedly, some readers will, at least occasionally, wish such a fate for this book.

11. An extreme example of a fairly common view is the following statement made by the physicist Frank Tipler on CBS 11 TV (Dallas/Fort Worth) on May 9, 2007: "God is the cosmological singularity. I am not being blasphemous. I'm just following in the ancient tradition in saying that science puts the tenets of religion up to the experimental test, and we find that God exists."

12. Until Hawking came along, Vladimir Belinsky, Isaak Khalatnikov, and Evgeny Lifshitz thought they could prove the absence of singularities in generic space-time solutions of Einstein's equations, only to stand corrected by Hawking's more powerful mathematical methods. Nevertheless, their analysis is still being used to understand space-time close to singularities and represents an important contribution to our knowledge of general relativity.

13. Oscar Wilde, "De Profundis."

2. QUANTUM THEORY

1. This is a serious problem for the transporter beam process popularized by the series *Star Trek,* by which an object or a person is completely measured, annihilated, and reconstructed identically at a new place. The very first step, the complete measurement, is already impossible due to the uncertainty principle; the second step, annihilation, seems much easier. According to the inventors of the series, this problem is supposed to be solved by the so-called Heisenberg compensator, which dodges the uncertainty relation. Unfortunately, not much information exists about details: Asked how the Heisenberg compensator works, a technical adviser once replied, "It works very well, thank you."

2. There are other energies associated with spherical wave functions, but they have larger radii. The one closest to the proton is the stable one, for it cannot be compressed further. Trying to do so would bring us in conflict with the uncertainty relation, limiting the uncertainty achieved for position and velocity at the same time: Compressing the wave function reduces the position uncertainty, for we get to know the size of the wave better. The velocity uncertainty must change accordingly when the radius shrinks, for the electron must remain in a smaller region. The limit to uncertainties then limits the electron's orbits.

3. The term "Planck length" was coined later by John Wheeler, who recognized its role in quantum gravity.

3. AN INTERLUDE ON THE ROLE OF MATHEMATICS

1. Friedrich Nietzsche, *The Gay Science* (Stuttgart: Reclam, 2000), 237.

2. In *100 Years of Relativity—Space-Time Structure: Einstein and Beyond,* edited by Abhay Ashtekar (Singapore: *World Scientific,* 2005), 18.

3. There is even a "theory" so mysterious that not even its name is known, except for its first letter. Indeed, this letter is often interpreted as standing for "magic" or "mysterious," but sometimes also for words such as "membrane," or even as being a mirror image of the first letter of its originator's surname, reflected at a horizontal axis.

4. Although the mathematical degree varies, theories in this sense exist not only in physics but also in other areas of science. A well-known example is biological evolution, which, based on principles such as selection and mutation, can explain the richness of the species of life on Earth as well as many specific details.

4. QUANTUM GRAVITY

1. In some crucial investigations, they were joined by Ted Jacobson.

2. These approximations are called *lattice gauge theories.* When he started

to work on loop quantum gravity, Smolin already had extensive experience in a version called Hamiltonian lattice gauge theory, which undoubtedly helped him make the application to gravity. A similar non-approximate framework, but not for gravity, had earlier been developed by Rodolfo Gambini and Antoni Trias—initially unbeknownst to Rovelli and Smolin.

3. The first important step was performed with another senior mathematical physicist, Chris Isham. Soon afterward, several results were obtained with Jerzy Lewandowski, Don Marolf, José Mourão, and Thomas Thiemann. Independently, Rodolfo Gambini and Jorge Pullin worked out alternative mathematical methods.

4. It also turned out that a reformulation of Ashtekar's original reformulation of general relativity was necessary, introduced by Fernando Barbero in 1995.

5. Several anecdotes illustrate this observation, best of all Lee Smolin's after-dinner recollection voiced at AbhayFest, a celebration of Ashtekar's sixtieth birthday in 2009, saying that his contact with Ashtekar taught him the difference between ideas and results.

6. Wilson loops are named after Kenneth Wilson, who introduced them for more general purposes in quantum field theories unrelated to gravity.

7. For instance, in work by Thiemann with Kristina Giesel.

8. There was, in fact, a good reason for his initial hesitation: He thought he could do better and had developed his own way of averaging symmetrical states. When he gave a seminar on this method, I pointed out that it would always give completely vanishing states: The symmetry would not only wash out structures, in this case it would completely remove everything. Ashtekar immediately realized that I was right, and after this experience he began to support loop quantum cosmology. In this whole process, he showed greatness, combined with a rare character trait that I have learned to respect highly, despite some often annoying side effects: the formation of and receptivity to opposition and criticism are essential in the pursuit of truth, and the more vigorous the criticism the better, unless they are empty and vain. Scientific striving can give rise to fierce, sometimes emotional combat. But this is usually constructive, for it quickly builds up theories, weeds out mistakes, and suggests new ideas. And after the final debate is over, one is left with a piece of knowledge agreed upon by all involved.

9. Pawlowski's newly developed computer programs were especially crucial to this research.

10. Everyone can participate in the data analysis thanks to the program Einstein@Home (http://einstein.phys.uwm.edu). This is a screen saver using a computer's idle time for an analysis of observational data. Currently, about 100,000 computers in many countries participate, providing gravitational research with welcome computational capacities.

11. Friedrich Schiller, *Was heisst und zu welchem Ende studiert man Universalgeschichte?* published in *Der Teutsche Merkur,* 1789 (pp. 127–128).

12. Not all descriptions of the universe follow the top-down view. As we will see in a later discussion of the uniqueness of solutions (chapter 9), loop quantum cosmology, for instance, can provide fundamental conditions for the form of its solutions without referring to the present state of affairs. It thus has more predictive power, but is also more daring: Predictions about the current state are made, and then must hold their own compared to observations.

13. Everyone loves a superpower that swoops in swiftly and with overwhelming force to save the day. Such a neoconservative view—ignoring many quantum aspects in spite of better knowledge—is often voiced in cosmology, suggesting that a single repulsive force would be sufficient to exorcise the threatening singularity and liberate the whole universe. Unfortunately, proclaiming solutions based on the consideration of only a single mechanism too often ignores the realities on the ground: In the real world, things are much less tidy, and different forces are at play. In the early universe, not only do we have the repulsive contribution to the gravitational force, but other, ubiquitous quantum forces are also relevant in such a dense, hot, messy phase. To understand clearly what is going on, detailed properties of the whole state must be known. How exactly the repulsive force of quantum gravity resolves the big bang singularity thus remains to be explored—but at a fundamental level, based on the behavior of energy in discrete space-times, it is clear that it does so in some way.

5. OBSERVATIONAL COSMOLOGY

1. The observable universe at that time had just about one-billionth of its current volume.

2. Penzias and Wilson had not developed their microwave antenna with the aim of cosmological measurements in mind; the discovery was connected to cosmology only in retrospect. Other researchers had independently planned similar experiments but were scooped by Penzias and Wilson. Among these researchers was David Wilkinson, a name to be met again in the context of the satellite WMAP, which currently provides the most precise data about cosmic background radiation.

3. At any given time, the combination of all observations can show us only a finite part of the universe. What if the part we have seen so far is just a tidy special region in a much more irregular universe? This possibility can never be ruled out; cosmological predictions of the extremely long-term kind can escape this conundrum only with extra assumptions or the postulation of general principles.

4. For some time, Reines and Cowan had even played with the thought of constructing the detector near an exploding nuclear bomb.

5. One could just as well call the present acceleration "inflation" instead of dark energy. But since these are two apparently distinct phenomena, they should be separated conceptually. And because accelerated expansion in

the early universe was historically investigated first, it had already claimed the name "inflation."

6. For many researchers in cosmology, inflation obtained from quantum gravity shows strong appeal. In loop quantum cosmology, I published the first humble (and, in hindsight, probably wrong) possibility in 2003, meant as an initial indication not to be taken too seriously. Much later, I learned that this article was the main stimulus for Parampreet Singh, who now is among the active researchers in loop quantum cosmology, to enter the field.

7. Why unequal amounts of matter and antimatter currently exist is an open problem. These two ingredients must either have been separated from each other, or have emerged in different amounts owing to some asymmetry in their production. Neither of these possibilities can currently be supported completely by known physical processes, as we will briefly discuss later in this chapter.

8. What we call matter and what antimatter is, of course, only convention. Matter is what we normally see, and it has a dual relationship with antimatter produced at high energies. Since we do not know the cause for the dominance of matter, there is no feature distinguishing it physically; it is distinguished only by the apparently random fact that it dominates our cosmos.

9. Employing this mechanism for tests of quantum gravity was first proposed by Giovanni Amelino-Camelia, John Ellis, Nikolaos Mavromatos, Dimitri Nanopoulos, and Subir Sarkar in 1997. Some of them developed specific models and estimates for time delays in certain nonstandard (so-called noncritical) versions of string theory; the standard, critical string theory is not expected to give rise to detectable time delays because it retains the classical structure of space-time to a high degree. In loop quantum gravity, such models were provided by Rodolfo Gambini and Jorge Pullin in 1998 for photons, and by Jorge Alfaro, Hugo Morales-Técotl, and Luis Urrutia in 1999 for neutrinos. With recent detections of high-energy gamma particles from very distant bursts, observations are now approaching sensitivities by which some quantum gravity theories can be tested and their free parameters constrained.

10. Gamma ray bursts were first discovered by the Vela satellite of the U.S. military, which was deployed to detect putative Soviet nuclear explosions in space.

11. The motion of all the stars, however, is not confined to a fixed plane. A 3-D animation can be viewed at http://www.mpe.mpg.de/ir/GC/index.php.

6. BLACK HOLES

1. Passages in italics are fictitious and are meant to illustrate which speculations the mathematical foundation of modern physics forbids that might otherwise be possible.

2. For rotating black holes, which are not rotationally symmetric but are typical in the cosmos, there are nevertheless new phenomena. For instance, according to the Penrose process, objects passing near a rotating black hole can be accelerated at the expense of the black hole's rotation rate; by stopping the rotation, one could in principle gain energy corresponding to up to 30 percent of the black hole mass. This is an enormous amount, as one can see if one considers that nuclear power relies on the transformation of a much lower percentage of the far smaller nuclear fuel mass.

3. Or drown, after several more years of global warming.

4. Unruh's suggestion started with a harmless metaphorical illustration in a colloquium talk on black holes. By now, actual experiments are being performed in different versions.

5. Which is by no means to say that investigations of black body radiation cannot lead to important insights. Just Planck's studies or the black body radiation of the cosmic microwave background need be mentioned here.

6. Here, initial contributions came, in different collaborations, from Kirill Krasnov, Carlo Rovelli, Abhay Ashtekar, John Baez, Alejandro Corichi, Romesh Kaul, and Parthasarathi Majumdar.

7. In parallel, similar investigations were undertaken by Leonardo Modesto as well as Viqar Husain and Oliver Winkler. More recently, Christian Böhmer and Kevin Vandersloot as well as Rodolfo Gambini and Jorge Pullin have turned to these questions.

7. THE ARROW OF TIME

1. An application of this principle is more realistic, but much more difficult, than that of ideal mathematical time; this is best expressed in the words of Peter Bergmann, who in the 1960s took the first steps toward this issue in relativity: "Such a question can, we are assured, always be answered from a sufficient set of initial data, though the performance of this task may call for considerable mathematical agility." The theoretical description still remains to be completed; in recent times it has been pushed forward by, for instance, Carlo Rovelli and, building on his work, Bianca Dittrich. Julian Barbour also continues to tackle such questions.

8. COSMOGONY

1. Of interest is the usual picture with one healthy and one broken tusk as in figure 33, symbolizing the combination of perfection and chaos in the real world; see also chapter 11.

2. Pre-Socratic fragments are translated from the German quotations in *Philosophie von Platon bis Nietzsche*, Digitale Bibliothek, 3rd ed., vol. 2, (Berlin: Directmedia, 2002), and Hermann Diels, *Die Fragmente der Vor-*

sokratiker: Griechisch und Deutsch, 4th ed., vols. 1 and 2 (Berlin: Weid-mannsche Buchhandlung, 1922), respectively.

3. At least this was the case in the early, innocent years of loop quantum cosmology. With promising results and higher stakes, the situation has changed: It has already been shown that singularities can be avoided; now the question is how space-time looks. This is a different issue, requiring a handle on more details that are easy to misinterpret. For instance, there were some discussions of the grand "quantum nature of the big bang," in blissful ignorance of the fact that the model used as well as quantum states analyzed in this context were very special—too special, as it turned out, to reliably draw detailed conclusions about the big bang. Another example is the claim that "loop quantum cosmos are never singular," a statement that tells it all about how contrived the contents are: The word "cosmos" is singular, but by being juxtaposed with the verb form "are," it is artificially made plural. In such cases, when the situation was scruti-nized more closely, initial claims had to be scaled back. Normally, this is a healthy procedure; first promises draw interest and focus research. It becomes problematic only when claims, by being overblown, spread widely but then are debunked only quietly.

4. The model was later evaluated further by Ellis with Jeff Murugan and Christos Tsagas.

5. In a different version, this solution was repeated later by Luca Parisi, Marco Bruni, Roy Maartens, and Kevin Vandersloot.

6. A more complicated possibility is that the universe might have come into existence some finite time ago according to a classical concept of time, yet always have existed in the sense of quantum evolution. Resolving this paradox requires an understanding of the subtle difference between clas-sical and quantum time and, eventually, a complete understanding of time itself. Such issues, which are fraught with not only technical but also con-ceptual problems, are still being worked out. At present, it is indeed con-ceivable that loop quantum gravity in general might give rise to such a chimerical view, in which both cyclic and linear features are present, as already indicated by cosmic forgetfulness; independently, those types of models could resemble a spontaneous inflationary initiation of the uni-verse in the sense proposed by Sean Carroll and Jennifer Chen. Funda-mentally, however, in loop quantum cosmology there always seems to be a quantum universe before what classical gravity would see as the singu-larity.

7. Please refer to note 1 for chapter 6.

10. THEORY OF EVERYTHING?

1. "That which tempts one to regard all philosophers half-distrustfully and half-mockingly is not the oft-repeated discovery of how innocent they

are—how often and easily they make mistakes and lose their way, in short, how childish and childlike they are—but that there is not enough honest dealing with them, whereas they all raise a loud and virtuous outcry when the problem of truthfulness is even hinted at in the remotest manner. They all pose as though their real opinions had been discovered and attained through the self-evolving of a cold, pure, divinely indifferent dialectic (in contrast to all sorts of mystics, who, fairer and foolisher, talk of 'inspiration'), whereas, in fact, a prejudiced proposition, idea, or 'suggestion,' which is generally their heart's desire abstracted and refined, is defended by them with arguments sought out after the event. They are all advocates who do not wish to be regarded as such, generally astute defenders, also, of their prejudices, which they dub 'truths'—and *very* far from having the courage of the conscience that bravely admits this, exactly this, to itself, very far from having the good taste of the courage that goes so far as to let this be understood, be it to warn friend or foe, be it in cheerful confidence and self-ridicule." (Friedrich Nietzsche, *Beyond Good and Evil*). Examples can be found also in quantum gravity. Quantum cosmology in particular, a field of comparatively undemanding mathematics with considerably higher stakes, has seen fearless struggle. Fortunately, in physics there is the conscience of experimental observations, powerful enough to command an end to overboiling theoretical heart's desires.

2. Physicists are hangmen: Our most noble, though unpopular, job is to kill those theories that have no right to exist. But most physicists are sentimental; they often fall in love with their victims. And the task only gets more complicated if a condemned theory happens to be one's own brainchild. The cute and charming ones, particularly, can easily survive well past their allotted expiration date.

3. Friedrich Nietzsche, *On the Genealogy of Morality*, Preface, §6.

11. THE LIMITS OF SCIENCE AND THE NOBILITY OF NATURE

1. Science, like so much else, works best in a democratic setting. How well this is realized in a subfield depends to a large extent on the ethics of its leaders. Some are seen as gods, some want to act like despots; there is but little room between the kings revered and kings reviled. Judging for a particular field requires a detailed view behind the scenes, usually available only to its practitioners. For loop quantum gravity, I can safely say that while it is certainly not perfect, on average at least it comes close to the ideal: There is a fearless tyrant, to be sure, but the rest are a bunch of anarchists. A vibrant community it is.

2. Friedrich Nietzsche, *On the Genealogy of Morality*.

INDEX

A NOTE ABOUT THE AUTHOR

Martin Bojowald is an associate professor of physics at Pennsylvania State University's Institute for Gravitation and the Cosmos. Originally from Germany, he now resides in Pennsylvania.

A NOTE ON THE TYPE

This book was set in Garamond, a typeface originally designed by the famous Parisian type cutter Claude Garamond (ca. 1480–1561).

Claude Garamond is one of the most famous type designers in printing history. His distinguished romans and italics first appeared in Opera Ciceronis in 1543–44.

COMPOSED BY
North Market Street Graphics,
Lancaster, Pennsylvania

PRINTED AND BOUND BY
Berryville Graphics,
Berryville, Virginia

DESIGNED BY
Iris Weinstein